ORGANIZING NATURE

Turning Canada's Ecosystems into Resources

Organizing Nature

Turning Canada's Ecosystems into Resources

ALICE COHEN AND ANDREW BIRO

UNIVERSITY OF TORONTO PRESS
Toronto Buffalo London

Toronto Buffalo London
utorontopress.com
Printed and bound by CPI Group (UK) Ltd, Croydon, CR0 4YY

ISBN 978-1-4875-9485-5 (cloth)
ISBN 978-1-4875-9484-8 (paper)
ISBN 978-1-4875-9486-2 (EPUB)
ISBN 978-1-4875-9487-9 (PDF)

Library and Archives Canada Cataloguing in Publication

Title: Organizing nature : turning Canada's ecosystems into resources / Alice Cohen and Andrew Biro.
Names: Cohen, Alice (Geographer), author. | Biro, Andrew, 1969–, author.
Description: Includes bibliographical references and index.
Identifiers: Canadiana (print) 20230168922 | Canadiana (ebook) 20230169007 | ISBN 9781487594848 (paper) | ISBN 9781487594855 (cloth) | ISBN 9781487594879 (PDF) | ISBN 9781487594862 (EPUB)
Subjects: LCSH: Natural resources – Social aspects – Canada. | LCSH: Nature – Social aspects – Canada. | LCSH: Environmental management – Canada. | LCSH: Canada – Environmental conditions.
Classification: LCC HC113.5 .C64 2023 | DDC 333.70971 – dc23

Cover design: Lara Minja
Cover images: iStock.com/ulimi; iStock.com/setory; GaudiLab/Shutterstock.com

We welcome comments and suggestions regarding any aspect of our publications – please feel free to contact us at news@utorontopress.com or visit us at utorontopress.com.

We wish to acknowledge the land on which the University of Toronto Press operates. This land is the traditional territory of the Wendat, the Anishnaabeg, the Haudenosaunee, the Métis, and the Mississaugas of the Credit First Nation.

University of Toronto Press acknowledges the financial support of the Government of Canada and the Ontario Arts Council, an agency of the Government of Ontario, for its publishing activities.

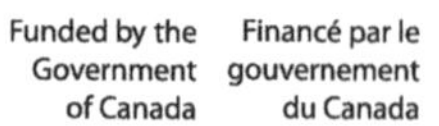

Contents

Illustrations and Tables

Illustrations

Tables

Maps

Big Ideas in Small Boxes

Acknowledgements

This book has two authors, but it was made possible by a village of support. First, it is important to recognize that the institution where we both work (Acadia University) and the town where we both live (Wolfville) are located in Mi'kma'ki, the ancestral and unceded territory of the Mi'kmaq. We acknowledge the Treaties of Peace and Friendship and thank the Mi'kmaq people for their generosity in sharing their homeland with us. This land acknowledgement is a small step in recognizing the significance of the relationship between Mi'kmaq and settler communities, and we hope to contribute positively to this relationship in this book and beyond.

Most of the book was initially drafted pre-COVID, with the two of us writing in the same physical space. Thanks to the many Acadia colleagues who encouraged us as we worked in room 210 of the Beveridge Arts Centre (the Politics and Philosophy Departments' lounge) and to the staff at Just US! Coffee in Wolfville. After March 2020, much of our work moved online. Thanks to the staff of Acadia's Technology Services Department for helping us to set up and navigate a virtual workspace.

In undertaking this work, we were also supported by funding from the Acadia University Research Fund. This allowed us to receive very helpful feedback on early versions of this work from colleagues at academic conferences. Here, particular thanks are due to Andrea Olive for her comments on an initial version of the project as a whole at the meeting of the Canadian Political Science Association in 2017 and to Jennifer Lawrence for her comments on Chapter 6 at the meeting of the Western Political Science Association in 2019. The funds also allowed us to hire Charles Douglas and Nina Polletti, who provided valuable research and editorial assistance, and Eric Leinberger, who produced the maps.

The approach taken in this book was given a test run in Alice's Politics of the Environment course in 2018–19; thanks to the students in that course for their feedback. Special thanks also to Kaela Biro for reading the whole manuscript with the eyes of an undergraduate student and to Erin Patterson for answering our copyright questions.

The editorial staff at the University of Toronto Press have been a pleasure to work with: Mat Buntin, Jennifer DiDomenico, Marilyn McCormack, Stephen Jones, Rebecca Duce, Mary Lui, and Kathie Porta Baker. Thanks also to Kris Gies for the initial email that gave us the idea for this book. Three anonymous reviewers solicited by the University of Toronto Press provided a careful read of the text and many helpful suggestions for improving the manuscript.

Since this project started, we have been through a maternity leave (Alice), a global pandemic, and a strike, among other events and actual or potential disruptions too numerous to list. Each of us is thankful to the other for maintaining our more or less regular writing meetings that allowed us to see this through to the end. Special thanks to our partners (Jamie Sedgwick and Lisa Speigel) for their support, including making it possible to chip away at this project while working from home during COVID lockdowns, and to our children (Sam and Molly; Kaela and Nathan) for their curiosity about this project.

Abbreviations

CBS	Canadian Blood Services
CDWQG	Canadian Drinking Water Quality Guidelines
CITES	Convention on the International Trade in Endangered Species of Wild Fauna and Flora
CMHC	Canada Mortgage and Housing Corporation
CO_2	carbon dioxide
CPR	Canadian Pacific Railway
DEVCO	Cape Breton Development Corporation
DOSCO	Dominion Steel and Coal Corporation
DFO	Department of Fisheries and Oceans
ECCC	Environment and Climate Change Canada
EROEI	energy return on energy invested
FOC	Fisheries and Oceans Canada
FPAC	Forest Products Association of Canada
FSC	Forest Stewardship Council
GDP	gross domestic product
GHG	greenhouse gas
IFF	Indigenous food fishery
IP	intended parent
MOU	memorandum of understanding
MSY	maximum sustainable yield
NAFO	North Atlantic Fisheries Organization
NCC	Nature Conservancy of Canada

NGO	non-governmental organization
OECD	Organisation for Economic Co-operation and Development
SQ	Sûreté du Québec (Quebec provincial police force)
SYSCO	Sydney Steel Corporation
TEK	traditional ecological knowledge
WCD	World Commission on Dams
WHO	World Health Organization

CHAPTER ONE

Introduction

We begin with an ad for Molson Canadian beer. The 60-second ad is called "Made from Canada" (Molson Canada, 2010) and begins, "When you think about Canadians, you might ask yourself, why are we the way we are?" It then goes on to answer, "It's this land that shapes us," including the fact that "we have more square feet of awesomeness per person than any other nation on Earth." The visuals that represent that awesomeness throughout the ad are almost exclusively expansive footage of the big outdoors, and the accompanying narration similarly refers to "lakes, mountains, forests, rivers, and streams" as "the best backyard in the world, and we get out there every chance we get." In short, the ad identifies Canada with wilderness, or unspoiled nature, and being Canadian as having a love for nature.

Cultural artifacts such as ads, art, and television spots provide glimpses into the knotty relationships between people and place in ways that are not always visible through formal government pronouncements. This is not to say that governmental institutions and policies are not important – they are – but rather that it is also important to broaden our field of vision to see environmental politics as encompassing more than just what governments or environmental groups do. Widening our analytical focus allows us to think about the relationship between people and place as a two-way street between people and landscape: in our view, environmental politics is about how people shape the land and how the land shapes them.

But let's get back to the beer ad. The ad seems to be saying that the Canadian environment shapes Canadians to be nature lovers. Particular kinds of claims about environmental politics are being made here: claims that might seem more controversial and worth contesting if they were made in a political debate rather than a beer ad. The ad hints at a kind of environmental determinism, that is, the idea that who you are depends on where you are from. There is also an assumption that there is one single definition of Canadian identity, one way to be Canadian (can you even be Canadian if you don't like hockey?). Are we reading too much political content into an ad? Probably, but we also think it is a useful starting point for this book because it provides a glimpse of the kinds of small moments that, together, make up the mosaic of Canadian identities as they relate to place. Like all beer ads, the point of it is to sell beer. But it is telling that the ad's creators thought they could sell a lot of beer to Canadians with this kind of appeal to a particular version of national identity. In this ad, representations of nature ("lakes, mountains, forests, rivers, and streams") are a symbolic resource that is used to encourage the extraction, processing, and consumption of material resources (water, barley, and hops, and glass or aluminium to contain it). This consumption, in turn, shapes the flow of yet other resources, including more than US$3 billion to the Denver, Colorado–headquartered Molson Coors Brewing Company in 2010, the year that the ad aired.

Moreover, we suggest that the beer ad only tells half a story. It says that the "land shapes us"; it says nothing about how Canadians have altered their environment. Indeed, the story of how Canadians are shaped by the landscape and the story of how they have shaped the landscape seem to run parallel to one another; never intersecting, but rather providing two quite different views of the relationship between people and place.

That these relationships are almost always framed as one-way streets underscores the idea of the nature–society binary: the idea that people and nature are two separate and distinct things or that people and society are here and nature is there. One of our key arguments is that this binary is both present and problematic in Canada. We also emphasize that both *nature* and *society* are loaded terms with no single, widely accepted definition. For example, Canadian society is not a singular thing: both authors of this book are descendants of

European settlers who arrived in a Canada that was and is home to Indigenous populations who had lived here for many thousands of years. And, of course, Canadian society contains many, many heterogeneities: regions, languages, ethnicities, urban–rural populations, and so on.

Similarly, nature is not a simple category. Canada has a colonial legacy in which a vast wilderness plays a leading role as something to be conquered, something to be used as a site of resource extraction, and something to be used as a recreational space.

So, the nature–society binary is problematic in two ways: first, because nature and society are constantly making and remaking each other, and second, because neither nature nor society is a homogeneous thing. The Molson ad suggests that an emblematic Canadian experience is set at a lake, in the mountains, on the coast, in the rocky landscape of the Canadian Shield, on the expansiveness of the prairies, or even just outdoors in frigid winter weather ("mon pays, c'est l'hiver," as Quebecois singer Gilles Vigneault famously sang). This imagined setting is distinctly at odds with the setting in which most Canadians find themselves most of the time. Eighty per cent of Canadians live in an urban environment. Virtually all Canadians' time at school is spent indoors, as is their work time (for most), and even much of their leisure time, and the way of life of most if not all Canadians is dependent, at least indirectly, on the revenues of extractive industries.

The contrast between an imagined wilderness and life in a highly managed environment sets up a binary that, we suggest, allows Canadians to embody a peculiar paradox: cultural infatuation with the natural landscape (see also A. Wilson, 1991) and economic dependence on the remaking and extraction of resources from it. The sustained effort to build and maintain this binary is a key theme throughout this book, and one that shapes the approach that we take to understanding environmental politics. The idea that society and nature are radically distinct is not the result of particular government policies, nor can we attribute its authorship to specific actors. It is an idea that suffuses many Canadians' way of thinking about and acting on the world around us, and it is an idea that is learned from and reinforced by a multitude of sources, from the operating principles of key social institutions to beer commercials. This in turn means that we take a somewhat unconventional approach to understanding environmental politics: one that looks at politics beyond policy, beyond the government institutions and regulations that shape what actions humans can (or cannot) undertake on the landscape. Instead, as we explain in the remainder of this introductory chapter, we explore politics beyond policy; examine not just how resources are governed but also why they are governed that way; explain the transformation of the non-human world from ecosystem components to resources; and explicate the channels through which these transformations are made possible.

1.1 FROM HOW TO WHY

Many excellent books explain *how* Canadian environments are managed. Our focus in this book is on *why* resources are governed in the ways that they are. To answer the "why" question, we need to expand our focus beyond policies and governmental institutions.

For example, understanding *how* water is governed includes understanding which levels of government are responsible for what dimensions of water governance, the processes by which transboundary waters are negotiated and managed, the mechanisms for enforcing water quality standards across the country, and so on. Conversely, understanding *why* water is governed the way it is raises different questions: Why do so many government agencies play a role in water governance? Why is water sometimes framed as a resource and sometimes framed as part of the environment? How is water transformed from H_2O into an object of governance? Why do some communities live with decades-long boil-water advisories while others have some of the best drinking water in the world? We suggest that addressing these questions is a necessary component of understanding environmental politics in Canada. Moreover, we argue that the artificial separation of nature from society – reproduced in Canada's governance institutions where nature is governed separately from resources – enables Canadians to have an economy based on resource extraction and processing while simultaneously holding a national identity that celebrates their collective connection to nature in the form of wilderness.

1.2 FROM ECOSYSTEM COMPONENTS TO RESOURCES

A key focus of this book is the transformation of disparate objects, things, or beings into resources (things that people can use instrumentally or things that can be manipulated to serve their ends). Although it is a clunky term, we occasionally use the word *resourcification* to emphasize that there is a process that needs to happen for things to become resources, or objects that can be treated in a particular way. Trees only become timber, for example, when they are cut down and processed. Forests are ecosystems unless they become resources. Resourcification, or resource thinking, is common in modern Western societies: some would argue that it has become people's default way of conceptualizing and interacting with the world around them. So it is worth pausing to emphasize that this is not the only way for people to think about and engage with the world.

The title of this book, *Organizing Nature,* can be interpreted in two ways. In one sense, it is about how humans organize nature and turn it into resources. This organizing takes work and often encounters resistance, and it is a large part of the story of how Canada became the country that it is today. (As we show, this is not necessarily a story with a happy ending, nor one in which those doing the organizing are unambiguous heroes.) In another sense, the title is about how nature has organized Canada: once things have become resources, they can reshape physical and social landscapes. For example, the deployment of lumber built railroads, and the deployment of money derived from the sale of oil and gas built Calgary into a city of more than a million people. Other kinds of organizing are less material: the creation of environmental departments within governments, subsidies to encourage or discourage particular behaviours, education curricula, hiking clubs, and on and on.

For something to become a resource, though, it needs to be conceptualized as an individual thing, something that can be removed or disembedded from its context or environment. Resources need to be extracted, physically moved from one place to another, literally put to another use. At the same time, this extraction is only possible if resources can be abstracted – that is, imagined as separate from their current context. What do we mean by this? First, resource thinking relies on abstract categories that make one individual thing commensurate with another: logs or board-feet of lumber as opposed to a particular tree whose identity is tied to a particular location or set of relations with other beings (e.g., the oak tree in the backyard that we hung the swing on). Second, resource thinking relies on a way of imagining the universe in terms of individual, isolatable beings, as opposed to a web of relationships. At least some of these relationships are physical: ecosystem components are place based, whereas resources have been mobilized or have become ageographical. If one compares this resource thinking with the first of Barry Commoner's (1971/2020) four laws of ecology ("everything is connected to everything else"; p. 33),[1] then one can see how resource thinking may be inherently anti-ecological. To think of something as a resource is to deny or forget how its connections to everything else are what make it what it is. The extractions and abstractions are colonial and are antithetical to many Indigenous perspectives wherein people, place, and natural object are inextricably intertwined. It is not surprising, then, that what we call the resourcification of the Canadian landscape is part of the colonial process of nation building.

Of course, because human beings (like all other living beings) need to transform their environment to survive, to some extent they need to think of the things around them as manipulatable objects. But where, and how firmly, should the line be drawn between objects to be manipulated for one's own purposes and subjects whose autonomy deserves to be respected?

The whole idea of a nature–society binary rests on the idea that there is such a thing as nature and such a thing as society, but these are complicated terms with long histories. Many Indigenous ontologies, for example, understand non-human animals and some landscape features such as rivers or mountains as persons or beings with spiritual significance equal to (or perhaps even greater than) what is accorded to humans. For example, Mi'kmaw scholar Margaret Robinson (2013) explains that "Mi'kmaq legends view humanity and animal life as being on a continuum, spiritually and physically. Animals speak, are able to change into humans, and some humans marry these

1 Commoner's (1971/2020) four laws are (1) everything is connected to everything else, (2) everything must go somewhere, (3) nature knows best, and (4) there is no such thing as a free lunch (pp. 29–43).

shapeshifting creatures and raise animal children" (pp. 191–2). She also goes on to say that Mi'kmaq "stories characterize animals as independent people with rights, wills and freedom" (pp. 191–2).

In spelling out his idea for a "land ethic" – an ethically appropriate way to think about and interact with the world around one – American ecologist Aldo Leopold (1966) wrote that "a land ethic changes the role of *Homo sapiens* from conqueror of the land-community to plain member and citizen of it. It implies respect for his fellow-members, and also respect for the community as such" (p. 240). Leopold knew that that this was an uphill battle in modern Western society (he was writing in the mid-twentieth-century United States) and that such an ethic was dependent on seeing land as "not merely soil" (p. 253) and on "the extension of the social conscience from people to land" (p. 246). Taking this a step further, Soren Larsen and Jay Johnson (2017) discuss a variety of cases in which places can be said to have agency, actively pulling human and non-human beings together into communities. Further examples in which legal rights have been afforded to features of the landscape such as rivers or mountains could also be provided (for a global overview, see D.R. Boyd, 2017). The point in all this is to establish that it requires considerable work to sustain (or change) a boundary between some who are deemed capable of making claims on others, and others who can only be acted on.

In the case study chapters (Chapters 3–8), we explore in more detail how the nature–society binary has been developed (and contested) in Canada, in the transformation of a variety of different objects from features of the landscape, or ecosystem components, to resources. The key point, however, is the existence, as a broadly accepted concept, of this line between nature and society and the binary thinking that it produces. Thus, the case study chapters provide a series of illustrations, showing how specific resources have been produced as such in Canada and with what consequences. At the same time, throughout the book we also make a more general argument about the nature–society binary as a hegemonic concept in the minds and lives of Canadians. As we look specifically at how forests are transformed into timber, fish into fisheries, and land into real estate, we are at the same time exploring the implications of looking at the world as mostly, and increasingly, consisting of resources.

1.3 POLITICS BEYOND POLICY

For us, politics is not limited to governments and their institutions, although it certainly includes them. To be sure, Canada's federal, provincial, territorial, municipal, and Indigenous governments play a critical role in determining the way in which humans interact with their surrounding environments. Although government policies do set rules within those frameworks, actors have considerable latitude in how they interact with their environment. Particularly in liberal societies such as Canada, governments do not play a dominant role in the economy and tend to preserve scope for individual choices. For example, more environmentally friendly products such as electric vehicles, fair trade coffee, or organic produce may all be available, but consumers are not mandated to buy them. What is more, although governments do set rules and try to enforce them, they are not always successful in doing so. Policies and laws are violated, unconsciously or consciously, both by those who feel that the rules are too stringent and by those who feel that environmental protections are not stringent enough. Thus, an account of environmental politics needs to be able to explain not just what the rules are, but also to what extent, and with what effects, the rules are (dis)obeyed.

Even if we stick with the frameworks provided by government-established policies and rules, the choices that individuals make are not completely uninfluenced by others. Power is exercised in many ways short of government regulation. The architecture of the choices people are presented with matters. One everyday example is how supermarkets can be physically designed so that some products are more likely to be purchased than others, by placing them at eye level or on the end caps at the ends of aisles. Advertising functions similarly, to induce demand, creating a desire for something that you did not even know you wanted. The power to be a choice architect or to mount an advertising campaign rarely rests directly with government. Instead, concentrations of economic power (owning a supermarket or media platform, or being able to mount a large-scale advertising campaign) are what allow some actors to exert considerable influence over others.

Of course, not all decisions are made by governments or economic calculations. For instance, many consumers willingly pay more for certified organic produce or for bottled water even when clean tap water is available, and many of the things that people value most deeply are things that they say they cannot put a price on, despite economists' sometimes elaborate attempts to do so. Although these considerations are psychological in the sense that they refer to mental states or ideas and values, it is important to stress that they are arrived at collectively rather than just individually. By living in Canadian society and participating in Canadian culture – watching beer commercials, among many other things – people can come to think of and value nature and Canada in these particular ways.

Finally, it is important to stress that this model of environmental behaviour being shaped by policy, economics, and culture (we specify six different channels in the next section and in Chapter 2) is a dynamic one. It may be tempting to think that government first sets the rules, then more or less powerful non-government actors use various economic or cultural means to influence others, and finally individuals make their choices, but the reality is more complex and recursive. Policy is dynamic, not established once and for all, and policy-makers are responsive to the kinds of economic and cultural considerations that emerge from individuals, who may in different moments think of themselves as consumers, workers, residents, citizens, and so forth. In this slow, complex, and messy process, Canadians have changed the landscape around them as biotic and abiotic things are converted into the raw resources that have built – and continue to build – Canada as it is today.

1.4 RESOURCIFICATION THROUGH SIX CHANNELS

Resourcification, or the transformation from ecosystem component to resource, requires movement of material objects and ideas. These movements, repeated over time, can produce well-worn grooves that make subsequent similar movements easier. The first time someone walks through a forest, it may be difficult going. As more and more people walk through it, the ground is packed down, branches are pushed aside or broken, and a path is created. After enough time and movement, the path, a product of human activity, appears to the people who walk on it to be an integral part of the forest itself.

Although paths are an appealing metaphor, we prefer to use the term *channel* to emphasize that the flows and movements we are discussing are not only flows and movements of material objects. They also include the flows and movements of ideas and discourses. Channels are used not only to move things from one place to another but also to circulate ideas, including expectations of culturally appropriate behaviour. Channels can be thought of not only as canals that provide a predictable path for water to flow through but also as YouTube channels that allow one to easily find videos that express certain kinds of ideas or forms of entertainment.

As noted earlier, in this book we are interested in explaining why certain kinds of environmental transformations happen: why the Canadian landscape has been (and is being) shaped in the ways that it has. Part of our answer to why Canada looks the ways it does today is because channels have been carved out – paths have been created – that make it easier or more difficult for certain kinds of transformations to happen. This is a phenomenon that social scientists call *path dependence* (Pierson, 2000). We emphasize that these channels work in two directions. Channels allow for human beings to transform the non-human landscape, and at the same time features of the non-human world help to determine or shape the channels. Together, this back and forth between people and place make the path seem perfectly commonsensical and natural.

Although the discussion of channels will at times get quite specific, one can also think of channels as nested one within another, as a small stream is part of a river basin or watershed, or a branch is part of a tree, which is part of a forest. The "governments" channel includes municipal, provincial and territorial, federal, and Indigenous governments, and although all provincial governments are similar in some respects, there are also differences between them.

In one way or another, all the channels in this book mediate between people and place. They create and solidify connections between Canadians and Canada in various ways. At least for our purposes in this book, the most significant transformation is from a world imagined as integrated ecosystem components to a world understood as a set of extractable resources.

1.5 BOOK OUTLINE AND COMMON THEMES

The rest of the book proceeds as follows. In Chapter 2, we outline the central theoretical foundations of the book. Specifically, we address each of the channels by explaining how we define it and what others have theorized about it and by identifying some key examples of its use. The ensuing chapters each give an example of resourcification, with a view to exposing and challenging the nature–society binary by making explicit the processes through which particular physical objects came to be classified as resources. In each case, after a brief introduction, we flesh out two or three historical or contemporary case studies and then identify the role(s) of the channels in facilitating those transformations.

Chapter 3 focuses on the transformation of fish to fisheries. Drawing on examples from British Columbia and Newfoundland, we make the case that this transformation was economic, governmental, cultural, and structural and that humans and fish have mutually influenced each other for thousands of years. In Chapter 4, we address the channels through which forests become timber and the physical, regulatory, economic, demographic, and cultural shifts that facilitate the transition. This case of resourcification also involves a more recent reconceptualization of intact forests as resources for tourism or climate stabilization. In the latter case, it may be more profitable to not cut down trees, but in both cases trees are understood as a resource, and we explore the ways in which the channels have facilitated that transition. In Chapter 5, we turn to carbon in its various forms: coal in Nova Scotia, bitumen in Alberta, and natural gas in a variety of locations across the country. Here, we address how the physical properties of fossil fuels – namely, the fact that coal is labour intensive to extract and bitumen and gas are capital intensive to extract – shaped how each was understood and developed as a resource. In addition, we explore the complex and mutual co-organization of people and place because coal towns and, more recently, the oil sands city of Fort McMurray, have developed their own identities and political character.

In Chapter 6, which concerns water, we take a conceptual turn from ecosystem components that themselves become resources to ecosystem components that facilitate extraction. In many ways, water is not just a resource in itself, but its availability and characteristics have also enabled the extraction of other resources: forestry, mining, fisheries, and agriculture. We also focus on the cultural dimensions of the resource because water holds an especially iconic place in the collective cultural imagination and in Indigenous cultures. In this chapter, we also highlight the complex institutional landscape of environmental governance: water is notoriously difficult to govern, in part because it affects nearly every sector of society. Similarly, the resourcification of land, addressed in Chapter 7, is central to the collective cultural imagination. Also like water, the significance of land as resource goes beyond the land itself. Land as resource also enables other extraction activities. We explore this phenomenon using three examples. First, we address soil and the ways through which the outermost layer of the earth's crust has become a resource for intense agricultural activity. Second, we look at land as a symbol – specifically, the national parks system and the role it has played in fuelling particular kinds of national identity. Third, we look at land as space – that is, as real estate. The idea that humans can buy, sell, and own land is a relatively new one in the grand scheme of human history, and it is one that is central to the development of mining and forestry enterprises.

Finally, Chapter 8 looks at the resourcification of an entirely different category of ecosystem component: those parts and functions of bodies from which profit can be derived. We start this chapter with an exploration of the commodification of non-human, non-agricultural animals, from pets to wild fish and game to protected areas, each of which is a resource in its own way. In the latter part of the chapter, we look at human bodies and how blood, plasma, and surrogacy have all been framed as resources in various ways; we also explore the ways in which various human attributes such as age, language, and so on are seen as resources in Canada's points-based immigration system.

Despite their different areas of focus, some common themes can be drawn from the chapters in this book. We encourage you to think of these common themes as "organizing nature glasses" that you can look through: they are structured ideas that can help you make sense of complicated issues. These ideas are used throughout the book in service of our central argument: that people and place are in a constant process of making and remaking each other and that the commodification of ecosystem components has been a central component of that relationship.

The first theme is commodification – that is, assigning monetary value to something so that it can be bought or sold. In the transformation from forest to timber, for example, tree trunks are given a monetary value. When forests are commodified as carbon sinks, their existence as live trees also carries a monetary value based on their size, age, ability to store carbon, and so on. Similarly, a fish swimming in the ocean as part of the ecosystem has no monetary value until someone catches it to sell or trade or thinks that they or someone else will be able to do so. Commodification is a key moment in the transition of something from an ecosystem component to a resource. Although commodification is part of resourcification, the latter is something broader.

A second theme or concept that arises frequently throughout the book is the idea of Indigenous dispossession – the taking of land and culture from Indigenous Peoples. This dispossession is not a relic of the past; it is an ongoing and active process. One way that Indigenous dispossession is sustained is through Indigenous erasure: the recurring and false idea that Canada was a vast and uninhabited wilderness when European settlers arrived. This idea features prominently in Canadian culture – think of famous paintings or songs featuring the Canadian wilderness – and shapes the relationship between people and place in very colonial ways. For Indigenous people whose ancestors were in fact here and living in thriving, complex, and sophisticated societies at the time of European arrival, the recurring images of a pre-colonial wilderness free from human presence erases family and cultural histories as if they never happened. For immigrants and their descendants (i.e., everyone who is not Indigenous), it tells a story about conquering nature – about the need to eke out a hardscrabble existence in the unforgiving wilderness. Although there was, of course, struggle and hard work, there is also significant evidence to suggest that European settlers were able to survive only because of the kindness and generosity of the Indigenous people they encountered. Yet, the stubbornly persistent imagined history of Canada as a vast and untouched wilderness remains central to resourcification.

A third concept is the idea of the existence of a nature–society binary, that is, the idea that nature and humans are two entirely separate things and that humans can exist outside of nature in order to master it. This idea has rightly been debunked many times, and we continue to do so here by highlighting the ways in which nature and society are deeply interconnected, with each continually reshaping the other: humans are always active participants in complex and dynamic ecosystems, and they are always responding to ecosystem changes, in terms of both individual behaviour and societal organization. When forests are logged for timber, the new planted forests are much less diverse in species and age and create entirely different kinds of habitats for plant and animal life; when bee populations collapse, humans develop the tools and inducements to get some people to pollinate crops by hand; when dams and waterways are built, fish must find new ways to travel; when nutrition guidelines change, human consumption of particular foods increases or decreases, and agricultural practices change in step; when certain resources are in high demand, people move to particular places to participate in its extraction and processing, thereby establishing new communities, changing not only the landscape but also the lives of the people who inhabit it. In no case do humans stand completely outside of nature, able to control or shape it without being shaped by it in turn. This false binary comes up in nearly every chapter of the book, and it is something you cannot unsee in the media (in beer commercials and elsewhere) once you start watching for it.

So, how do ecosystem components become resources? Through the channels, all of which are enabled by and infused with themes of commodification, Indigenous dispossession, and the false nature–society binary. Moreover, the channels help to explain how people and place are a set of interlocking gears driven in part by colonial nationalism and the role of resourcification in building Canadian national identity. Throughout the book, we show how resource thinking was instrumental in creating Canada as it is today. If the dominance of resource thinking globally has led to ecologically unsustainable ways of living, as we think it has, then thinking about how people might live on the northern half of North America without relying so deeply on this kind of thinking is an urgent task. An important first step for this task is appreciating how, and how much, this way of thinking has shaped the world and oneself.

DISCUSSION QUESTIONS

1 This chapter begins by discussing a popular text (beer advertisement) from the early 2010s that links Canadian identity with the Canadian landscape. What are some more contemporary examples of Canadian texts (ads, TV shows or films, songs, etc.) that make similar links? How are Canadian landscape and identity portrayed? How does the text work to connect them?
2 Think about an environmental problem that is important to you; it could be a specific pollution problem in your local community or a global problem such as climate change or biodiversity loss. To what extent is this problem driven by resource thinking?

CHAPTER TWO

Channels: From Ecosystem Components to Resources

2.1 INTRODUCTION

One of us once hiked with a forester friend who would walk along the trail pointing at old-growth trees, saying, "There's $10,000, and there's $10,000, and there's another $10,000!" Such exclamations can only be made within the scaffolding of a society in which resource thinking is dominant: those old-growth trees have financial value because there are companies with forestry licences and with the equipment and labour to cut down trees, remove them from the forest, and transport them to customers willing to pay for them. That company, those licences, that equipment, and those customers exist because there are governments that issue licences, people willing to work in forest communities, and customers willing to pay for the resulting wood. This embedded network is resourcification in practice.

So how do ecosystem components become resources? It does not happen overnight, and it works very much like a series of interlocking gears: once you turn one small wheel, the whole system moves. But what are the wheels, how are they connected, and who gets to turn them? In this chapter, we explain the resourcification process by envisioning channels through which ecosystem components become resources – that is, pathways through which the process takes place. In the following sections, we identify some of those channels and give examples of some of the ways they have served to facilitate resource

Table 2.1. Channels for Organizing Ecosystem Components into Resources

Channel	Domain (what happens in this channel)	Roles in resource thinking	Representative example
Governments	Laws and policies	Regulating extraction; nation building and dispossession	Fisheries and Oceans Canada; see Chapter 3
Communities	Group identity and social life	Geographic communities: proximate to extraction sites, contribute to path dependency Non-geographic communities: identity creation, education, and advocacy	Gold rush in Dawson City; see Chapter 7
Built environments	Landscape transformation	Normalizing the nature–society binary (cities–wilderness); infrastructure construction	Fort McMurray; see Chapter 5
Culture and ideas	Stories and ways of thinking and feeling	Amplifying settler world views; creating and unifying disparate communities; facilitating identity politics	"Log Driver's Waltz"; see Chapter 4
Economies	Distribution mechanisms	Commodification; enabling consumer identities; driving labour mobility	Water privatization; see Chapter 6
Bodies and identities	Individual lived experience	Providing the labour of extraction (or enabling extraction); mobility to resource communities; identities in nation building; bodies or parts of bodies as resources	Blood and plasma donation; see Chapter 8

thinking. In subsequent chapters, we look at specific case studies, each of which draws on the channels introduced here. In this chapter, we describe six channels (Table 2.1), although the distinctions between them are not always firm.

A word of warning to readers: we provide an overview of each of the channels separately in this chapter; however, in the case study chapters that come afterward, discussion of the channels is woven together in sections titled "Channels in Action." There, the names of the channels appear in boldface for easier reference, but careful readers (and careful observers of the real world!) will no doubt note that in practice the channels are woven together.

2.2 GOVERNMENTS

When thinking of environmental problems and their solutions, there is often a tendency to think of governments first – and with good reason: governments play a significant role in mediating the relationships between people and place. It is governments – federal, provincial and territorial, municipal, and Indigenous – that regulate human activity with respect to the environment: they issue permits to logging and mining companies, create and enforce pollution legislation, define and oversee environmental impact assessments, regulate vehicle emissions, issue fishing and hunting permits, offer green construction rebates, and on and on. Although the central argument of this book is that considering environmental politics beyond governments is critical for seeing the whole picture, understanding the role governments play is still an important component of that picture.

The key point we want to emphasize about governments in this book is that they organize their environment, and many governments are organized to reflect features of the Canadian landscape. Understanding this relationship as a two-way street is important, because environmental analyses typically focus more or less exclusively on governing human behaviour with respect to the environment. For example, there are debates over which level of government is responsible for (or can profit from) resource extraction (e.g., Hill & Harrison, 2006) and sophisticated and thoughtful analyses of why particular governments are motivated – or unmotivated – to interpret the constitutional division of powers in particular ways (Harrison, 1996). These analyses are invaluable in understanding the protection of, or extraction from, the Canadian landscape, but they tell one little about how the physical attributes of the resources have also driven the organization of society, including governments.

In telling the story of Canadian resources, a common narrative is that the environment exists (more or less) as an inert object to be governed. What we are suggesting is that the environment and government institutions exist in a mutually co-constitutive relationship: environments shape government institutions and the way they function, and vice versa. As a concrete example, Nova Scotia has a Department of Fisheries and Aquaculture, and Saskatchewan does not, because the latter does not have a significant commercial fishery. Similarly, British Columbia has a Ministry of Forests, Lands, Natural Resource Operations and Rural Development that reflects the long and complex history of forestry in that province, and for several years Manitoba had a minister of water stewardship, who reflected that province's physical location downstream from several major watercourses. In all of these cases, governments have organized themselves in ways that – in some respect – reflect the resource realities of their respective jurisdictions. In these examples, the resources have (passively) done the organizing of government. In turn, this organizing was facilitated by an understanding of these ecosystem components (fish, forests, water) as resources in the first place.

2.3 COMMUNITIES

We use the word *community* in two ways. The first is in terms of geographic community – that is, groups of people living or working in the same place (see Box 2.1). There is considerable overlap between geographic communities and governments because many geographic communities are represented by governments: for example, the communities of Montreal and Iqaluit are represented by their respective municipal governments. Other geographic communities (neighbourhoods or regions), however, might not have analogous governments.

The second kind of community is non-geographic – that is, a group of people who are not necessarily physically proximate but who have shared interests. Examples of this category include things like environmental communities and fishing communities, or groups not directly related to environmental issues, such as foodie communities, LGBTQ+ communities, or scientific communities. Although non-geographic communities predate the existence of the internet, social media have facilitated their development and growth. The distinction between geographic and non-geographic communities may be helpful in some cases and aids in highlighting some trends, as we discuss later. However, as with many of the categorizations in this book, the reality is not always clear-cut, and there is often overlap between geographic and non-geographic communities: think of, for example, the Banff skiing community or the Yellowknife arts community.

BIG IDEAS IN SMALL BOXES

BOX 2.1. RESOURCE TOWNS

Canada provides many examples of geographic communities embedded in the resourcification process. Each community's story is unique, but they share a common storyline: a valuable resource was discovered, a community was established, individual and collective economic futures were at stake, and identities were forged. In many cases, this arc eventually comes to a point at which the resource is no longer valuable. Perhaps the resource in question is mined until it is gone; perhaps it becomes too expensive or too difficult to extract; or perhaps a better, cheaper, safer, or cleaner alternative is found. In any case, the community must then grapple with what it means to exist in the absence of its central organizing force. In all these cases, the physical and political landscape is transformed.

The discovery of the resource means not only finding the object but also developing the technological and social means to turn it into an extractible resource. Trees in New Brunswick became a valuable resource in the context of shipbuilding technologies and European mercantile empires. Bitumen in northern Alberta became a valuable resource in the context of oil extraction and processing technologies, pipelines, and a car-centric North American culture. Both of these (and many others) became resources in a legal regime that created and enforced property rights and reframed ecosystem components as resources that were extractible, movable, and exchangeable.

With the preconditions for extraction in place, the resource in question acts like a magnet, drawing people together who make individual and communal lives in a particular place, organized around the extraction of a particular resource. Consider the case of Cape Breton County in Nova Scotia, where 250 million tonnes of coal were extracted between 1863 and 1976 (Gillis, 1978). Journalist Alison Auld (2001) described the cultural impact of coal mining when the last coal mine in Cape Breton was closed: "For many families and towns that rose up around the area's 12 bustling collieries, it's almost impossible to imagine a Cape Breton without mining. It has so deeply shaped the island psyche and soul with its own language, superstitions, unique brand of camaraderie and defiant spirit it seems impossible that it would disappear" (paras. 8–9).

The attachment of specific resources to particular geographic communities is seen even in the names of sports teams, such as the Edmonton Oilers and the Brandon Wheat Kings.

What happens to a resource town after the resource that organized it is exhausted? Resource exhaustion does not necessarily mean the complete physical depletion of the thing itself. For example, the population of Fort McMurray, Alberta, grows and shrinks not on the basis of the amount of bitumen remaining, but on the price of oil. Some resource towns grow and diversify to the point where they are no longer as reliant on the resource around which they were organized: Ottawa has long outgrown its previous identity as the lumber town of Bytown. Others reorganize the resources and the ways in which they are used, for example by turning former extractive industry resources into sites for tourism and leisure. Still others shrink dramatically, such as Dawson City, Yukon, which grew to 40,000 people during the gold rush at the turn of the twentieth century but has had a population of fewer than 5,000 people for more than a century. Others are abandoned completely, like Pine Point, Northwest Territory (Weaver, 2015).

Non-geographic communities, too, provide examples of resources organizing society. Consider, for example, movements such as Fridays for the Future or the Canadian Association of Petroleum Producers and how they bring together geographically disparate actors for a common cause in ways that shape, and are shaped by, ecosystem features and functions.

Fishing communities provide an example of this mutual co-evolution between people and place. In terms of geographic communities, there are certainly many places in Canada that were organized around the presence of fish and that have a significant proportion of their residents engaged in fishing. These geographic communities depend on the extraction of fish. However, as fish, fishers, and fisheries have all become increasingly organized at the provincial and even national and global scales, and as these fishing communities have been increasingly integrated into geographically broader communities, these local (geographic) fishing communities' cultural distinctiveness – the distinctiveness of their ways of doing things, as opposed to their being in a distinct or specific location – has diminished. At the same time, this organization of the fishing industry at higher scales has helped to produce a non-geographic fishing community in which people are connected by occupation rather than geographic location. Provincial organizations such as the Newfoundland Inshore Fisheries Association or national ones such as the Canadian Council of Professional Fish Harvesters work to advance the collective interests of their members. These organizations act more like lobby groups than grassroots communities. As with many of the resources that we discuss in this book, the industrialization of fisheries shows that as the extraction of a resource becomes increasingly organized, the people directly involved in that extraction also become increasingly organized, interacting in more impersonal, formalized, and rationalized ways.

2.4 BUILT ENVIRONMENTS

The Government of Canada defines the built environment as including "the buildings, parks, schools, road systems, and other infrastructure that we encounter in our daily lives" (Public Health Agency of Canada, 2014, para. 1). In other words, the built environment is those parts of the world that are constructed by humans for human use. Of course, the distinction between built environment and other environment is fuzzy. For one thing, the built environment is made of things that people think of as part of nature: water pipelines in Canadian cities were made of hollowed-out trees until the early 1900s, and homes are made of wood, stone, metal, and other materials derived from the ecosystem. Ecosystem components, in other words, are understood as resources and physically organized in different ways. These built environments can affect their surrounding areas: cities, which have concentrations of surfaces that attract and retain heat, experience higher temperatures than the surrounding countryside (the urban heat island effect), hillside construction can affect erosion patterns, impermeable surfaces such as concrete and asphalt prevent water from being absorbed into the ground, dams affect fish migration, and so on.

Of course, built environments are nothing new. Since time immemorial, Indigenous Peoples have lived in and modified disparate environments across North America. Examples include Haida in British Columbia's temperate rainforest; Mi'kmaq on the shores of Atlantic Canada; Assiniboine in and around what is now Winnipeg; Huron-Wendat, Haudenosaunee, and Anishinaabe Peoples in what is now Toronto; and Blackfoot in the prairies. The materials available in these locales – such as fish, buffalo, berries and other plants, and trees – as well as access to transportation allowing for trade with other communities, have fed, clothed, and sheltered Indigenous Peoples for generations. European settlers were likewise drawn to resource-rich areas, both for access to resources and to trade routes. Because the Canadian project was so heavily reliant on revenues from resource extraction, settlement was both actively encouraged and economically possible in those areas in which ecosystem components could readily be converted to resources and exploited for economic gain. Towns such as Fort McMurray, Alberta; Dawson City, Yukon Territory; Cobalt, Ontario; and Miramichi, New Brunswick, have become part of the physical landscape in the cycle of making and remaking both people and their environments.

Although non-Indigenous settlers and their descendants dramatically reshaped parts of the physical landscape, it is important to note that the landscape they encountered was not an uninhabited or pristine wilderness (Cronon, 1995). Indigenous Peoples were present on and in relation to the land when settlers arrived. In his foundational essay "The Trouble With Wilderness," environmental historian William Cronon (1995) traces the deep irony in the narrative of the European settler arriving in an "uninhabited wilderness." Cronon details the removal of "Indians" from their homes to create national parks in the United States:

> The myth of the wilderness as "virgin" uninhabited land had always been especially cruel when seen from the perspective of the Indians who had once called that land home. Now they were forced to move elsewhere, with the result that tourists could safely enjoy the illusion that they were seeing their nation in its pristine, original state.... Meanwhile, its original inhabitants were kept out by dint of force, their earlier uses of the land redefined as inappropriate or even illegal. To this day, for instance, the Blackfeet continue to be accused of "poaching" on the lands of Glacier National Park that originally belonged to them.... The removal of Indians to create an "uninhabited wilderness" – uninhabited as never before in the human history of the place – reminds us just how invented, just how constructed, the American wilderness really is. (p. 79)

Although Cronon was writing about national parks in the United States, the myth of the uninhabited wilderness was also pervasive in Canada. Indeed, the nascent Canadian government provided inducements for Europeans to settle on the prairies. Free land and money were given to settlers to work the "virgin soil" (Figure 2.1) from which Indigenous people had been dispossessed.

As historian William Denevan (1992) notes, however, "The roots of the pristine myth lie in part with early observers unaware of human impacts that may be obvious to scholars today" (p. 379); they also lie in part in the fact that by the time the interior lands of the North American continent began to be explored in 1750–1850, the Indigenous population had already been reduced by as much as 90 per cent through war and disease. These "honest mistakes" were convenient for the colonizers, who had an interest in seeing the land as uninhabited, a clean slate ready for the arrival of European settlers as part of the project of nation building across the continent. Indigenous Peoples were dispossessed from and had their presence erased from the land.

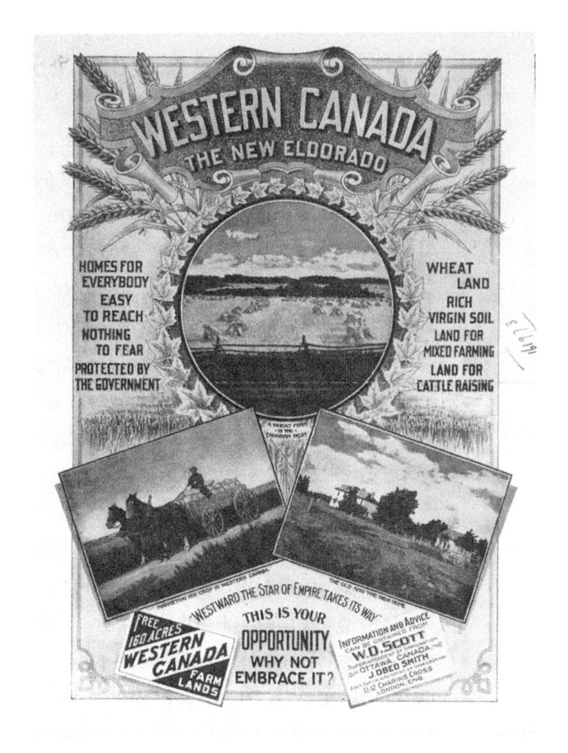

Figure 2.1. Western Canada – The New Eldorado, ca. 1890–1920
Source: Library and Archives Canada/Department of Employment and Immigration fonds/e010779321_s1, https://central.bac-lac.gc.ca/.redirect?app=fonandcol&id=2945432&lang=eng.

Although Indigenous Peoples did transform their environments, the landscape transformations that derive from European modernity, in Canada and around the world, far exceed the scope of previous transformations, to the point at which earth system scientists speak of the current epoch as one dominated by human influence (see Box 2.2). It is true that ecosystems, non-human species, and the features of the natural landscape all play a role in shaping the kinds of lives that people live, but so too do urban and other built environments. And although one might think of Canada and Canadian identity in terms of iconic nature – the Rocky Mountains or polar bears – the reality is that most Canadians spend far more time in cities than on mountains and looking at electronic screens (often featuring content produced outside of Canadian borders) rather than at unmediated nature. Most of the time, for most people, it is the built environment that surrounds (environs) them.

This reshaped environment both facilitates and is facilitated by the resource thinking described in Chapter 1, wherein the colonial gaze sees the landscape as composed of extractible resources and empty land to be settled by Europeans (Figure 2.2). Seeing the world as a set of resources makes possible the construction of a vast built environment and the lives people live as Canadians within it.

BIG IDEAS IN SMALL BOXES

BOX 2.2. THE ANTHROPOCENE

The term *Anthropocene* is used by some scholars to describe the current geological epoch. Earlier geological epochs, for example, the Holocene, Pleistocene, and Paleocene, are defined by specific geological or biotic events: the movement of continents, the flourishing and extinction of specific groups of species (e.g., dinosaurs, large mammals, winged insects), or periods of glaciation.

What makes the Anthropocene different from the most recent Holocene era is that it is defined by human – rather than geological – activity. Indeed, as Mauser (2006) writes, "The term Anthropocene has been suggested to mark an era in which the human impact on the Earth System has become a recognisable force" (p. 3). Debates exist about the specific start date of this human-driven era. Paul Crutzen (2006), for example, suggests that it may have begun in the latter part of the eighteenth century, coinciding with the invention of the steam engine in 1782 and reflecting what ice cores tell us are the beginning of the growth of greenhouse gas (GHG) emissions. Writing a century and a half ago but pointing to a similar timeline, "Stoppani in 1873 rated mankind's activities as a 'new telluric force which in power and universality may be compared to the greater forces of earth'" (as quoted in Crutzen, 2006, p. 13). A more recent timeline posits that the Anthropocene began around 1950; the International Geological Congress suggests that the epoch's beginning should be aligned with nuclear bomb tests in the middle of the twentieth century, although it notes that "an array of other signals, including plastic pollution, soot from power stations, concrete, and even the bones left by the global proliferation of the domestic chicken were now under consideration" (Carrington, 2016, para. 2). The key point here is that almost every dimension of the earth's functioning is affected by humans in some way. In particular, the past two centuries have seen "human-driven alterations of i) the biological fabric of the earth, ii) the stocks and flows of major elements in the planetary machinery such as nitrogen, carbon, phosphorus, and silicon; and iii) the energy balance at the Earth's surface" (Steffen et al., 2007, p. 614).

Despite some debate about the era's official start date, scientists recognize that we are now living in an era in which the earth's systems are shaped significantly by human activity and its effects. The question now, is how – or whether – this fact will change the way people govern themselves. Steffen et al. (2011) suggest that the business-as-usual approach will be catastrophic and that what is fundamentally needed is a change in governance structures that will foster planetary stewardship with the aim of keeping the Earth's systems in a Holocene-like state. The dramatic biophysical and political changes driven by this reality have "opened up serious questions that go to the heart of the social and human sciences" (Delanty & Mota, 2017, p. 32) – questions about environmental and social justice, about human–environment relations, about the role of governments and economies in this new reality, and about how to make the "right" decisions moving forward.

Figure 2.2. "It's Mine"
Source: "It's Mine! Canada – The Right Land for the Right Man. Canadian National Railways – The Right Way!'" (ca. 1920–35; CU12926299) by Canadian National Railways. Courtesy of Libraries and Cultural Resources Digital Collections, University of Calgary, https://digitalcollections.ucalgary.ca/asset-management/2R3BF1OAC8PVT.

2.5 CULTURE AND IDEAS

Culture and ideas is a broad channel that includes stories that people tell, real or imagined, through a range of media from books to TikTok videos. The Molson ad described in the introduction to this book is an excellent example of how the nature–society binary is mediated through this channel. The ad aims to summon feelings of national pride to encourage the watcher to purchase a particular beer. Why would a buyer be more likely to buy beer "made from Canada" than any other kind? More broadly, what is it about the idea of Canadian nature that encourages some kinds of environmental extraction and export (oil, forestry) but discourages others (water, blood products)? Answering these questions requires looking at the role of culture and ideas – that is, those parts of Canadian identities promoted through media, advertising, music, art, movies, and so on – in shaping the understanding of Canadian nature. For example, the famous "Heritage Minutes" vignettes are mentioned many times throughout the book, and in the forestry chapter we discuss the art of the Group of Seven in promoting a particular kind of Canadian landscape – one devoid of human activity – a *terra nullius* upon which the settler story can be written.

In seeking to characterize the role that the media play in contributing to the relationship between individuals and the state, famous Canadian communications theorist Marshall McLuhan (1964) argued that "technological [communication] media are staples or natural resources, exactly as are coal and cotton and oil.... Cotton and oil, like radio and TV, become 'fixed charges' on the entire psychic life of the community. And this pervasive fact creates the unique cultural flavor of any society" (p. 35). What does McLuhan mean by this? Communications media are required for the transmission of culture and ideas; they create a link (or relationship) between sender and receiver. If institutions are defined as (relatively) "permanent solutions to permanent problems" (Berger & Luckmann , 1967, pp. 69–70), then Canadian media institutions solve the permanent problem of conveying information (including information that entertains) to Canadians: media institutions allow them to know where to go if they want to find out who won an election or what the weather will be like tomorrow or if they want to listen to music or watch something entertaining. In answering these everyday questions and needs, relations are created between sender and receiver and between audience members who now share a common experience – an "imagined community" (Anderson, 1983), often despite a lack of personal contact or connection. Whether the information that is conveyed is presented as a factual representation (news) or a fictional one (entertainment) or as a combination of these ("infotainment"), the stories that Canadians tell each other, and the media used to communicate those stories, shape how Canadians view each other and the world around them. Even with the rise of the internet and social media as global communication networks, the national-scaled broadcasting media that McLuhan mentions still exert an outsized influence on cultural life and mental frameworks.

Many cultural touchstones in Canada come from the Canadian Broadcasting Corporation (CBC), a broadcaster funded in part by the Canadian government. The CBC's mandate is described in the national *Broadcasting Act* (1991), which stipulates that the CBC "should provide radio and television services incorporating a wide range of programming that informs, enlightens and entertains" and that

> the programming provided by the Corporation should:
> 1 be predominantly and distinctively Canadian, reflect Canada and its regions to national and regional audiences, while serving the special needs of those regions,
> 2 actively contribute to the flow and exchange of cultural expression,
> 3 be in English and in French, reflecting the different needs and circumstances of each official language community, including the particular needs and circumstances of English and French linguistic minorities,
> 4 strive to be of equivalent quality in English and French,
> 5 contribute to shared national consciousness and identity,
> 6 be made available throughout Canada by the most appropriate and efficient means and as resources become available for the purpose, and
> 7 reflect the multicultural and multiracial nature of Canada. (s. 3)

Thus, one can see that the CBC is explicitly tasked with nation building (points 5 and 6), bringing Canadians together just as did physical infrastructure such as the railroad (see Box 7.2). It is also tasked with reinforcing a particular conception of what Canada is and who Canadians are: regionally diverse (point 1), committed to liberal ideals of free expression (point 2), bilingual (points 3 and 4), and multicultural and multiracial (point 7).

Aside from the CBC, most news media in Canada are part of the private sector, operated as for-profit businesses. In the private sector, media ownership in Canada is highly concentrated, with a handful of companies owning most newspapers and television stations. Postmedia owns a majority of the newspapers published in Canada, and only two cities – Toronto and Montreal – have two or more paid daily newspapers with different owners. Although norms and legal protections of press freedom in Canada are relatively strong, for decades ownership concentration has raised recurring concerns about who gets to tell what kinds of stories (Ross, 2018; Theckedath & Thomas, 2012). In New Brunswick, for example, pointed questions are often raised about the implications of the Irving family's massive forestry and oil and gas holdings, combined with their near-monopoly control of news outlets in the province (see, e.g., Deneault, 2019). Online news sources and social media have certainly added more diverse voices and perspectives to news stories, but declining ad revenues resulting from competition from internet giants such as Google and Facebook have also made it much more difficult to make a news business profitable.

Although different from the media institutions, universities are another important source for the generation and spread of ideas. They operate independently from government, although in Canada almost all universities are publicly funded to some extent. Universities play a distinctive role in mediating the people–place relationship in several ways. First, they play a central role in educating the public about both the physical and the cultural dimensions of the Canadian landscape, through natural science classes such as geology, biology, chemistry, and environmental science and through humanities and social science classes such as history, politics, economics, literature, geography, and the fine arts. These classes also teach what are considered (in) appropriate ways to interact with the landscape, from proper forest management techniques to best practices in environmental policy development. Second, universities reflect the Canadian landscape by focusing on expertise germane to their particular locale. For example, at last count there were eight universities in five provinces offering forestry degrees: all are in provinces in which forestry is a prominent part of the provincial economy (British Columbia, Alberta, Ontario, Quebec, New Brunswick). Similarly, universities in British Columbia, New Brunswick, Nova Scotia, Prince Edward Island, and Newfoundland and Labrador all offer programs in fisheries, aquaculture, or marine management.

Knowledges generated by and through this channel – everything from science to popular culture – produce, reflect, and shape tropes and stereotypes about what Canada is really like, who Canadians are, and how Canadians inhabit the landscapes that they occupy. These stories can be both descriptive (telling people what things are like) and normative (telling them how things should be), and they can homogenize what it means to be Canadian. They may claim that they capture the essence of people's relationship to their environment in a particular time and place. Classic examples of this include *Anne of Green Gables*, set in nineteenth-century Prince Edward Island, or *Degrassi*, set in late twentieth- and early twenty-first-century Toronto.

We are at pains to point out here that Canadian identity is not at all a homogeneous thing. It varies tremendously by region (and subregion; as anyone from an urban centre such as Toronto, Montreal, or Vancouver will tell you, each neighbourhood has its own distinct character), age, language, race, educational status, religious and cultural background, Indigenous status, gender, and so on. More important, any individual is part of all these categories at any given time – you can be Canadian and francophone and female and Indigenous, and those identities may not overlap entirely with those of your household, neighbourhood, or peer group. There are a near-infinite number of ways in which people identify themselves.

Unsurprisingly, given the colonial context in which we live, some cultures and ideas have had a louder voice than others in establishing what has become the dominant Canadian identity. During an election campaign debate in 2015, Stephen Harper referred to "old-stock Canadians," suggesting that some articulations of Canadian identity are more authentic than others. Consider what many might describe as an archetypal Canadian experience: a canoe trip. The idea of escaping from the city to watery nature in a kids' summer camp or a family camping trip is one that we return to. The canoe itself is an object powerfully articulated with Canadian identity: historian Pierre Berton is alleged to have said that "a Canadian is somebody who knows how to make love in a canoe." This connection between canoes and Canadian identity is perhaps not surprising, considering the vast number of waterways that dot the Canadian landscape. The canoe was a powerful symbol for the Hudson Bay Company, established in 1670 as a fur trading company. Objects and how they are used in the landscape are used to tell particular stories and reinforce ideas about Canadian identity. Canoes are, of course, not a European invention – they were used by Indigenous Peoples throughout the continent for centuries before Europeans' arrival. For a reversal of the colonial narrative, see the terrific four-minute stop-motion animation of "How to Steal a Canoe," by Michi Saagiig Nishnaabeg songwriter, scholar, and artist Leanne B. Simpson, which is about a young Nishnaabeg woman and older Nishnaabeg man who steal a canoe from a museum – although, of course, it is clear that the museum is the original thief.

The relationship between human society and natural environment is reciprocal. Canadian cultures and values are shaped by Canadian environments, and those cultures and values have shaped how Canadians have thought about, interacted with, and transformed those environments. Culture and ideas also have material impacts (and material objects and processes affect culture and ideas). Like the boundaries between society and environment, the boundary between ideal and material, or mental and physical, is at times fuzzy, porous, and shifting. Again, our identification of distinct channels is a matter of us trying to impose some order, for the sake of analytic convenience, onto a messy reality.

2.6 ECONOMIES

Human evolution has always depended to some extent on the availability of ecosystem components for food and shelter. Many Indigenous cultures – from the Gwich'in People in what is known as Yukon to the Mi'kmaq People in what is known as Nova Scotia – depended (and still depend) on things like fish, foraged foods, wood, and wildlife. At what point, however, does a forest become timber? When do fish become fish stocks? Arguably, it is when there is sufficient interest in paying for that thing; that is, when it becomes a commodity – a thing that can be bought and sold. It is important to note that interest on its own is not enough: there is a whole set of relationships (both material and discursive) that is required for interest to be translatable into monetary terms. That is, economies are a channel through which ecosystem components become resources, but they are not the only channel driving that process.

With that in mind, one way to begin thinking about economies as channels for resourcification is to start by noticing that despite being an advanced and wealthy society in a post-industrial world, much of Canada's formal (money-based, or monetized) economy still revolves around the physical extraction or harvesting of what are often called "natural resources." Canada has a resource-based economy or, to state it even more strongly, since the early colonial period Canada was created for, and continues to be maintained by, the organization of nature into resources. Economic historian Harold Innis (1930) used the biblical phrase "hewers of wood and drawers of water" to describe the Canadian economy. For Innis and his contemporaries, the roots of Canada's economic structure in the fur trade marked it as a "staples" economy, in which economic growth and development were driven by the extraction and export of staple commodities – goods that are important in consumers' everyday lives. This is a pattern that persists to this day, as the extraction, processing, and sale of certain ecosystem components remain some of the primary contributors to Canada's gross domestic product (GDP). In subsequent chapters, we show how physical and social landscapes have been organized to facilitate the extraction and sale of particular resources (i.e., fish stocks in Chapter 3, lumber in Chapter 4, and fossil fuels in Chapter 5). As these changes, which are designed to make extraction and sale easier, become institutionalized – for example, when the government creates a Department of Fisheries – it becomes more difficult to create, or perhaps even imagine, an economy that is not organized around those resources. If those resources are not renewable, or if they are extracted at unsustainable rates, or if the demand for that resource declines, then communities can be caught in a "staples trap" (Haley, 2011): institutionalized overreliance on a particular resource that inhibits the development of alternatives. Just as water flowing downhill carves out a path that makes subsequent flows easier, the organization of resources similarly produces channels that reinforce collective ways of thinking and behaving.

Here it is also important to note that many of the decisions about what gets extracted, produced, or transported are made by private-sector firms or individuals. The private sector of the economy (broadly defined as those parts of the economy not run by government and consisting of firms or organizations aiming to make a financial profit) shapes, and is shaped by, the landscape in important ways. First, many private-sector organizations make a profit by extracting, refining, and selling components of the environment. For example, timber companies cut down trees and sell wood and wood products within Canada and elsewhere; the fishing industry removes fish biomass from oceans and lakes to sell; oil companies extract bitumen from northern Alberta and ship it elsewhere for refining and sale; mining companies extract and sell metals and minerals; and agricultural activity converts forests and marshes to farms. In all these ways, the Canadian landscape is physically transformed by economic activity. The competition between private-sector actors can exacerbate the problems of the staples trap as the planning and coordination of investment becomes more difficult: the landscape is transformed with an eye to private, and often short-term, profit (Box 2.3).

These enterprises are also shaped by the Canadian landscape. The Molson beer advertisement discussed in Chapter 1 takes great pride in this: Molson proudly declares that its beer is literally "made from Canada." Many of the country's most profitable businesses – and the communities and related industries they support – would not exist in their current forms without the resources from which they derive profit. Moreover, the location and physical characteristics of particular environmental features shape the way businesses operate: where they are located, to what extent they process the resources they are extracting, who their main customers are (and where they are located), and so on. For example, much of the bitumen extracted from the oil sands in northern Alberta is refined elsewhere, and so it is transported by pipeline, road, and rail to refineries and then onward for sale. It is valuable because of technologies such as the internal combustion engine and the vast global material and cultural infrastructure that supports "automobility." The materiality of this resource matters: if the oil sands were near an ocean for easy transport by ship, or if bitumen were useful on its own, political debates about pipelines would be either nonexistent or entirely different. This example shows the iterative relationship between the landscape and the economy: the physical attributes of the landscape have led economic actors to be organized in particular ways. Particular kinds of infrastructure (e.g., pipelines, roads, rail) and technologies (e.g., steam extraction, refining plants) are required for extraction to be profitable, and that infrastructure, in turn, physically changes the landscape. In addition, the infrastructure changes the governance landscape, as, for example, the existence of pipelines creates a need for regulatory agencies such as the Canada Energy Regulator, a federal agency that regulates the more than 73,000 kilometres of pipeline that crosses provincial boundaries, and provincial authorities that regulate smaller pipelines in their respective provinces (Canada Energy Regulator, 2021).

We began this section by suggesting that the formal, money-based, or commodity economy is not all there is to economies. Often when one hears or reads about the economy, it is in abstract, numerical terms: economic growth figures, the national unemployment rate, interest rates, stock market indices, and so on. How the ups and downs of these numbers affect people's daily lives is not always clear. Jim Stanford (2008) begins his book *Economics for Everyone* by debunking the idea that economics is a mysterious realm understandable only by experts: "Economics is quite simply about how we work. What we produce. And how we distribute and ultimately use what we've produced. Economics is about who does what, who gets what, and what they do with it" (p. 1).

If one thinks about economics as "who does what, who gets what, and what do they do with it?" then one can see that economies encompass much more than monetary exchanges that can be reduced to statistical figures. Clearly, some of who does what, who gets what, and what they do with it is determined in the private sector, in markets where goods or services are bought and sold. Consider, for example, the issue of food. You can get locally grown organic food at a farmer's market, if you have the money to pay for it. How much

BIG IDEAS IN SMALL BOXES

BOX 2.3. THE SPATIAL FIX

The *spatial fix* is a term coined by geographer David Harvey (1982) to describe how capitalist economies deal with the problem of uneven geographical development. Harvey begins by noting that regional differences in both natural endowments and labour productivity lead to geographically uneven development. As a result, some sites are more attractive than others for investors, workers, or both. As the profitability of particular investments declines over time, investors begin to look for new opportunities (places) to invest. The same logic applies for workers faced with declining wages or conditions of employment and for policy-makers charged with facilitating economic development more generally: as existing resources dry up or become harder to access, capital will move to a new place to fix (solve) the problem of declining profitability, lower wages or unemployment, or slowing economic growth. *Fix* implies a resolution (as in "the problem is fixed") and attachment (capital becomes fixed to a particular point in space). More important, it also implies that the resolution is only temporary, as with a drug addict needing a fix (Harvey, 2001, p. 24).

This dynamic alternation between movement and fixity can be seen in all kinds of development in both the private and public sectors. Substantial public investments are made to open up new frontiers for development and to facilitate the transmission of goods, services, and information: highways, ports, rail lines, electricity-generating stations and transmission wires, and so on. However, particular features arise when investment decisions are dominated by private-sector actors: the flow of resources are uneven, and the fixes solve problems for some while causing problems for others. Private-sector actors may be in competition with each other and, because of that, have limited information and ability to coordinate. Perhaps more important, a tension exists between investing capital in particular places so that profits can be realized (clearing land and building a factory, hiring and training employees) and keeping capital fluid or mobile so that it can take advantage of new opportunities. Harvey (2001) summarizes this tension, or contradiction, for capital in this way: "It has to build a fixed space (or 'landscape') necessary for its own functioning at a certain point in its history only to have to destroy that space (and devalue much of the capital invested therein) at a later point in order to make way for a new 'spatial fix' (openings for fresh accumulation in new spaces and territories) at a later point in its history" (p. 25). Investment and resource flows are necessarily dynamic, moving from one place to another, reshaping landscapes by building up investments of fixed capital and then eventually abandoning them for greener pastures. The spatial fix involves the seeking out of new resources to displace those crises to other places or future times. Spatial fixes are often rational from the perspective of the individual (or corporation or even government) pursuing them, but irrational from the perspective of the socioecological system as a whole.

As we describe in more detail in Chapter 3, during the early years of European settlement of North America, a great deal of capital was invested in building fishing communities in what is now Atlantic Canada, making it possible to make profits from fishing. As fish stocks declined, and as other economic opportunities arose, the investment required to maintain those communities became increasingly difficult to justify economically. A similar dynamic can be observed in the forestry and fossil fuel sectors (see Chapters 4 and 5).

money you have, in turn, is determined by the labour market – how much an employer is willing to pay for the skills that you have and the work that you are willing to do. The price of food at that market, too, is a function of many decisions: how farmers are paying for the land, how much they are paying workers to work on the farm, the cost for them to rent a booth at the market, how much gas they used to get there, and innumerable other factors. Markets, however, are not the only places where those questions are resolved. States, or the public sector, are also sites where questions of production and distribution are decided. Taxation and labour market regulation, such as minimum wage laws, have an impact on how much people are paid for their work, as well as how much and what kinds of work people do, and some goods and services are allocated by government policy rather than by market transactions. It is important to note that questions of production and distribution are also decided within households, where work (cooking food, cleaning dishes, or weeding the garden) is often unwaged.

If one includes not just the public sector and households, but also all kinds of other private but non-market transactions (helping a friend move, lending a book to a neighbour, or volunteering at a food bank), then one can see that what is often called the economy (things produced by waged workers for a corporate firm to be sold on a market) is actually just the tip of the iceberg of economies (Gibson-Graham, 2006).

To summarize, throughout this book, when we talk about economies, we are talking about the interactions of markets, states, and households and other private non-market settings.

Transactions in those spaces are about who does what, who gets what, and what do they do with it, and these decisions all take place in particular environmental contexts, or ecosystems. But we also go further than putting economies into an environmental context, because professional economists also often speak about the economy as something completely separate from the non-human world – or, perhaps more precisely, they acknowledge non-human nature, but only as a specific (natural resource) sector. In this way of thinking, in this sector, resources are always already present, ready and waiting to be extracted. Thus, we join others in showing that this conceptual separation of economy (or society) and non-human nature is false and misleading, and we extend the conversation by showing how this false separation is produced: how ecosystem components must be extracted and abstracted to get to the point where they can be treated as resources. In short, economies – what they do and how people think about them – are a central channel through which resourcification occurs.

2.7 BODIES AND IDENTITIES

A final channel through which the nature–society binary is mediated and ecosystem components are converted to resources is bodies and identities. This may sound bizarre at first, but bodies – people's physical selves – are a key piece of the complex machinery addressed in this book. The relationship between bodies and politics more generally is the subject of significant work by social theorists, among them Michel Foucault (1977/1979) and Judith Butler (1993), both of whom argue in their own ways that human bodies are both political and politicized. Consider the example of the household as a site where economic decisions are made, as discussed earlier. In households, unlike in markets, things are often distributed without money as a mediator (e.g., parents let children live in their homes without paying rent). Unlike with states, where decisions and policies are impersonal (ideally, at least), decisions within households are highly personal: just because parents let their children live in their home rent free does not mean that they would let a stranger do so. Within households, relations are intimate rather than impersonal, but goods and services and work responsibilities are nonetheless distributed. Even so, things are not necessarily distributed equally, and power is still exercised. Within the household, bodily identities and relations can play important roles in determining how much agency can be exercised and what share of household resources can be claimed.

Meanwhile, ecological thinkers have sought to criticize anthropocentric (human-centred) views that assume that political life is reserved only for human beings (Donaldson & Kymlicka, 2013; Leopold, 1966). More recently, post-humanist and new materialist thinkers have blurred the sharply drawn distinctions between human and non-human, and living and non-living, beings (Braidotti, 2013; Coole & Frost, 2010). Rather than seeing human beings as distinctively "above" non-human nature, they argue that human and non-human worlds should instead be seen as mutually constituting a web of life (Moore, 2015). Although not identical, these projects are similar to the undermining of the nature–society binary that is a concern of this book. They allow people to see their embodied selves simultaneously as subjects with agency, as objects influenced and regulated by other humans, and as vectors that are acted upon, enabled, and constrained by non-human beings, from micro-organisms to global systems.

Outside of environmental concerns, there are concrete and familiar examples of human bodies as objects of regulation. Consider, for example, blood alcohol limits for driving. How people manage their own bodies also has ecological implications. And the production of food that Canadians choose to consume also has environmental effects: cows produce methane (a powerful GHG), so removing dairy and meat from the most recent Canadian food guidelines may nudge Canadians in the direction of more climate-friendly diets. Another example is a recent spike in blueberry consumption, driven by an interest in its antioxidant properties. Blueberry bushes are fertilized with nitrogen, which has effects on the soil and surrounding environment. Blueberries thus provide an example of how Canadians' emphasis on the health of their bodies shapes the way the ecosystem functions. And, of course, drinking water regulations sit at the intersection of human bodies, chemistry, and ecosystem function: protected drinking water areas are an example of a protected ecological landscape with a direct link to human bodies.

In addition to these material connections between human bodies and ecosystems, we also explore the ways in which Canadian bodies become representations of national identity. If that sounds strange, consider the following. As the examples in Chapter 1 show, many of the ways in which Canadian identities are asserted or constructed are related to the particular environments in which Canadians find themselves. Having access to "more square feet of awesomeness" – as claimed in the Molson ad discussed in Chapter 1 – apparently shapes who Canadians are. The relationship between national identity and territory, however, is not straightforward. The territory of Canada is itself diverse, and idealized constructions of the Canadian landscape often differ from the reality on the ground. The 49th parallel and 141st meridian (between Alaska and Yukon) were defined as boundaries through political processes – they did not simply appear out of nowhere. When the boundaries were drawn, they cut ties that existed across what is now the border: for example, the Salish nation on the West Coast was split in two by the boundary (Norman, 2012). Residents of Canada's border communities might feel a closer connection with the neighbours down the road, on the other side of the border, than to people living in Canada thousands of kilometres away. Indeed, national identity is complex and multifaceted. Some immigrants and their descendants might identify with another ethnicity in addition to being Canadian, for example, as Italian Canadians or Chinese Canadians. Quebec nationalism and assertions of Indigenous sovereignty coexist in different ways, sometimes uneasily, alongside or within Canadian national identity.

As mentioned earlier, people's identity is made up of more than just their national identity. People who share a common national identity are differentiated in many other ways. As with national identity, these other forms of identity also shape and are shaped by interactions with the environment. We discuss a few of the many other ways people identify themselves in the remainder of this section, emphasizing how they are relevant to struggles over the construction of nature in Canada.

Gender is often presented as one of the most foundational and natural components of a person's identity, connected with biological sex and announced at birth (or earlier). Although norms are shifting and gender fluidity is becoming widely recognized, the connection between gender identity and environment is a long-standing concern. In *The Death of Nature*, Carolyn Merchant (1980) showed how the scientific revolutions of the early modern era were based on, and reinforced, sexist assumptions about the nature of the surrounding world, justifying gender-based discrimination by naturalizing the male dominance then found in Western society. Whether the connection between women and nature is seen as biologically rooted and invariable (as Merchant saw it) or as culturally constructed (e.g., Sandilands, 1998), gender identity is a powerful lens through which resources are organized. The gender division of labour affects the kind of work individuals do, both within and outside the household, as well as the resources to which they have access. Gender roles and identities also more generally affect how people interact with their environment. Cara Daggett's (2018) concept of petro-masculinity, discussed in Chapter 5, is a particularly provocative example of this.

Like gender, race is often presented as an immutable and biological (natural) feature of a person's identity. However, the idea that human beings could be subdivided into biologically (as opposed to culturally) distinctive racial groups, easily recognizable through markers such as skin colour, is associated with the development of the large-scale transatlantic slave trade just a few centuries ago. The history of race as a biological category is thus closely tied with theories of racial superiority and inferiority (Jackson & Weidman, 2004, pp. 1–27). This in turn means that racial identity itself has continued to be a kind of resource that conditions access to environmental amenities and exposure to environmental risks and harms, in both obvious and subtle ways (Box 2.4). Although the environmental justice movement originated in the United States, and despite significant differences between Canadian and US history, institutional racism is a feature of Canadian society, too. Canada also has a legacy of environmental racism that affects both Black and Indigenous communities (Agyeman et al., 2010; Gosine & Teelucksingh, 2008; Waldron, 2018).

BIG IDEAS IN SMALL BOXES

BOX 2.4. ENVIRONMENTAL RACISM AND ENVIRONMENTAL JUSTICE

Starting in the 1970s in the United States, studies of waste disposal facilities and other environmental "bads" showed that the siting of these facilities had a racial dimension. As early as the 1980s, the US government recognized that hazardous waste facilities had a disproportionately negative effect on poor and minority communities. It is not surprising that these facilities are located in communities that have less access to economic and political resources: for both government and private-sector actors, putting potentially environmentally harmful or risky facilities in those communities is the path of least resistance (Bullard, 1990/2000). Robert Bullard (1990/2000) further explains that racial identity itself is a political resource that some can use to escape environmental harm or risk: "The differential residential amenities and land uses assigned to black and white residential areas cannot be explained by class alone. For example, poor whites and poor blacks do not have the same opportunities to 'vote with their feet.' Racial barriers to education, employment, and housing reduce mobility options available to the black underclass and the black middle class" (p. 6). Researchers and activists brought more attention to environmental racism as a structural feature of American society – that African Americans are "the wrong complexion for [environmental] protection" (Bullard & Wright, 2012). The recognition of commonalities among relatively small-scale social movements against locally unwanted land uses led to the organization of a national environmental justice movement. In 1991, at the First National People of Color Environmental Leadership Summit, delegates drafted and adopted 17 principles of environmental justice ("Principles of Environmental Justice," 1991). Three years later, in 1994, then-US President Bill Clinton issued Executive Order 12898 ("Federal Actions to Address Environmental Justice in Minority Populations and Low-Income Populations").

Since the 1990s, the environmental justice movement has developed in at least two ways. First, there is increasing recognition that the dimensions of environmental justice include not only the distribution of environmental risks and amenities (outcomes) but also recognition and participation in the process of environmental decision making (Schlosberg, 2009). Second, the movement has grown beyond the borders of the United States. Although the general concepts may be applicable globally, distinctive histories of racialization and inequality mean that environmental (in)justice will look different in different national contexts. Some examples of environmental justice analyses in Canada include Agyeman et al. (2010), King (2014), Teelucksingh et al. (2016), Waldron (2018), and Weibe (2017).

Indigeneity is a related but distinct form of identity. Indigenous Peoples have a distinctive connection with the land, grounded in cultural teachings (Indigenous knowledge) representing the distillation of thousands of years of careful observation of local environments. This observation, and the passing of its lessons down to future generations, was necessary for community survival. As we show in more detail in Chapter 7, Indigenous conceptions of land are quite different from dominant settler conceptions of land as inert and commodifiable (real estate). As many scholars have noted, the idea of land as property is a colonial idea that underpins settler accumulation and Indigenous dispossession.

Indigenous rights in Canada are evolving, although, critically, these legal rights are a result of work to amend and evolve the settler legal system rather than a meaningful recognition of Indigenous systems of justice as they evolved for thousands of years before European contact. For example, in Canada, Indigenous Peoples also have unique legal (nation-to-nation) relationships with the Canadian state, recognized in Section 35 of the Canadian Constitution (*Constitution Act*, 1982), but this constitution is a creation of the settler government. Similarly, much (although not all) of the territory of Canada is governed by legal treaties signed between Indigenous nations and the Crown (the Canadian state or, before that, the British state), but these treaties are based on maps that delineate boundaries and property – again, settler introductions. In some cases, the treaties recognize the sovereignty of the Crown over a particular territory (e.g., the numbered treaties in Western Canada), whereas in others nation-to-nation status is affirmed and Indigenous sovereignty remains unceded (e.g., the Peace and Friendship Treaties in Atlantic Canada). Significant areas (especially in British Columbia) are not governed by any treaty. In any case, both legally and culturally, Indigenous Peoples in Canada have a distinctive relationship with their environments, with the federal government, and with the provincial and territorial governments in their territory.

A final form of identity to discuss here is class. *Class* can be defined in various ways, but broadly it refers to access to economic resources (wealth or income), the social status or prestige that confers authority or privilege, or both. Marxists define class more specifically in terms of an individual's relationship to society's processes of production. Members of the capitalist class, who own and therefore are able to make decisions about society's means of production, are able to organize resources (open a mine or a factory, clear-cut a forest) in ways that others cannot, as we discussed earlier in the "Economies" section.

Class in the broader sense of wealth or income also matters in terms of environmental impact. In general, environmental impact is a function of income. That is, the wealthier one is, the bigger one's environmental footprint. This pattern is found in Canada and elsewhere, and not just at an individual level: high-income countries such as Canada have a higher environmental impact per capita than poorer countries, which is a central dimension of international environmental politics. However, sustainable or environmentally friendly consumption choices such as organic foods or electric cars are often more expensive, making a green consumer lifestyle available only to those with greater income or wealth.

Whether defined in terms of economic power or social status, class is also connected to existential security. To put it in simple terms, wealthier people are less worried about putting food on the table or having a roof over their heads. This in turn shapes how people think about ecosystems. For example, as a result of economic wealth, people's collective imaginations have shifted when it comes to "wilderness." Two hundred years ago, one would have to be brave to leave the relative security of urban Europe to come to what was thought to be the wild and unpopulated (although, of course, there were people here) continent of North America. This wilderness was seen as a dangerous place where one could easily freeze or starve to death. Nevertheless, settlers did it, because it was seen as a way out of poverty – a way to work toward a better life. In today's context, individuals will still move to new areas to seek work – Fort McMurray, Alberta, is a prime example – although the move is not loaded with the same baggage about risking life and limb to rough it in the wilderness. As American scholar William Cronon (1995) notes, the wilderness has transformed in people's collective imagination: it was once a scary and dangerous place; it is now an aesthetic resource, source of contemplation, or place for adventure. To avail oneself of the wilderness in this way requires money for things like travel, accommodation, gear, and so on – so much so that the argument has sometimes been made that environmentalism is only a concern of the wealthy elite. Of course, this critique is flawed in many ways (witness, e.g., the environmental justice and Indigenous rights movements), but it does point to the role that class identity can play in defining perspectives on the relationship between people and place.

2.8 SUMMARY AND CONCLUSIONS

In this chapter, we have described six channels: governments, communities, built environments, culture and ideas, economies, and bodies and identities. These channels each show the mutual reshaping of people and (non-human) environments. We return to these channels in each of the case study chapters that follow because they help to illuminate and clarify how each of these particular resources has shaped Canada and Canadians. More generally, throughout the book, we emphasize how human societies and their environments are in a constant cycle of making and remaking each other. An animating question for us is the one that environmental historian Donald Worster (1985) asked in his study of water in the American West: "How, in the remaking of nature, do we remake ourselves?" (p. 30).

In answering that question, we highlight two points that are woven into the book. The first is that the remaking of oneself through the remaking of nature happens whenever one takes particular bits of nature (ecosystem components) and puts them to specific uses. When people repeat that same conversion (i.e., from ecosystem component to resource) over and over, patterns emerge that shape where and how they live and interact with each other. The second point is that human beings are a part *of*, not apart *from*, nature. This is not to suggest that it is impossible to distinguish between people and their environment, or between human society and non-human nature. Rather, it is to say that these distinctions are not as fixed as they may sometimes seem. As we show most clearly in Chapter 8, it is not only non-human entities that are conceptualized as resources (particular ecosystem components that can be put to specific uses). These distinctions, between society and nature, actor and resource, people and environment, are always to some extent arbitrary and are never politically neutral. The channels discussed in this chapter contain concentrations of political and social power: people who occupy certain positions within institutions can speak authoritatively and get other people to do their bidding. Governments, communities, economies, and built environments are all structured to facilitate some kinds of social actions while discouraging others, and the shaping of bodies and identities, as well as culture and ideas, makes various behaviours seem desirable or natural, or not. As we describe in more detail in the chapters that follow, all of these channels are dependent on resources being organized in particular ways. There is a very real sense in which Canada as people know it would not exist if wood and metal had not been organized into a railroad line that extended from the Atlantic to the Pacific coast.

At the same time, this physical (re-)organization of the landscape itself depended on seeing and thinking about the world in certain ways: seeing a forest as potential railroad ties and seeing the earth as a potential mine.

This way of seeing, thinking about, and acting on the world – environment as resources – has dominated Canada's historical development and continues to loom large in contemporary environmental politics. However, this view is not, and has not ever been, uncontested. And even within this view, conflicts exist over who gets to use resources and for what purposes. In both general, abstract terms and specific policy and governance decisions, and through a range of different channels, the environment-as-resources view contends with alternatives that resist the nature–society binary and its characterization of nature as mere resources to be used for organizing human society. The results of these struggles over how nature is constructed – what has or has not been organized as a resource and what those resources have been organized for – matter: it is what has (re)shaped the physical as well as the social and cultural landscapes that have produced Canadian ways of life. The following case study chapters show how these struggles have played and continue to play out for several specific resources.

DISCUSSION QUESTIONS

1 Think about the physical space in which you are reading this book – university campus, home, library, coffee shop. How has the physical landscape around you been transformed by human action? What kinds of institutions, ideas, or technologies made those transformations possible?

2 This chapter outlined several examples of channels maintaining and reinforcing the three themes discussed at the end of Chapter 1 (commodification, Indigenous dispossession, and the nature–society binary). What groups or individuals are contesting this? How are they using the channels to resist processes of commodification, Indigenous dispossession, or the nature–society binary?

3 In this chapter, we described some of the ways in which government institutions perpetuate the nature–society binary and the dispossession of Indigenous Peoples. Can you think of ways in which governments can act to change these patterns? What might be the obstacles to that kind of action, and how might they be overcome?
4 Some writers suggest the possibility of a "good Anthropocene," where humans' technological capacities are used to engineer a flourishing planet. Do you think this is feasible? If yes, what changes would be needed for Earth-changing technologies to be used in a good way? If no, what are the limits to this kind of technological change?

CHAPTER THREE

From Fish to Fisheries

July 2, 1992: On one side of a hotel ballroom door, people are working furiously to insert the front two legs of a chair through the door's handles to keep the door closed. It is hard work, because the door is being repeatedly hammered from the outside by a group of angry fishermen holding one of their members' bodies by his shirt, using him as a battering ram to try to open the door. Inside the room, Federal Fisheries Minister John Crosbie has just made an announcement: effective immediately, there will be no more cod fishing in Newfoundland. With a single announcement, 35,000 Newfoundlanders are out of work – the largest single layoff event in Canadian history. The fishermen outside the room are trying to get in to express their rage with Crosbie, who then uttered the most famous line of his career: "I didn't take the fish from the goddamn water" (cited in Maher, 2020, para. 9). The moratorium was devastating for the province, which experienced high unemployment, poverty, and outmigration in the subsequent decade – not to mention the cultural and personal effects of the sudden end to a way of life. The moratorium demonstrated the extent to which people are dependent on resources for sustenance and money but also, in many cases, for a sense of identity and pride.

3.1 INTRODUCTION

This chapter explores the resourcification of fish and the channels through which fish are both organized and organizing. We start by examining the organization of fish (and fish as resources used to organize) in two distinct places and times: (1) changes to salmon fishing rights and structure in British Columbia and (2) the collapse of cod stocks in late-twentieth-century Newfoundland and Labrador. Historical and geographic differences allow us to discuss the role and evolution of federalism, extractive capitalism, and settler colonialism. In the second half of the chapter, we analyse these events as manifestations of the various channels through which fish have been organized as resources and fish resources have organized Canada and Canadians.

Governance systems for fish and fisheries are complex. The Constitution (*Constitution Act*, 1867, s. 91) gives the federal government the power to regulate "Sea Coast and Inland Fisheries," and the agency that leads on this is Fisheries and Oceans Canada (FOC), sometimes referred to as DFO, the abbreviation for its earlier name, the Department of Fisheries and Oceans. The most important piece of legislation is the federal *Fisheries Act* (1985), originally enacted in 1868 and amended several times since then. As Anthony Scott (1982) notes, however, except for the open ocean, provincial governments have considerable power: "the right to fish continued to be held by the provincial Crown in 'inland waters' and the 'foreshore'" (p. 790). Although governments may regulate the fisheries, like most other sectors of the economy in Canada, it is private individuals and businesses that do the fishing and that, within the framework of government regulations, have considerable discretion about whether, where, when, how, and how much to fish. Further complicating matters are distinctions in types of fishing licences issued by FOC: commercial, recreational, and Aboriginal, effectively creating three distinct fisheries, sometimes covering the same fish. As we show in this chapter, fishing rights have been at the heart of several struggles to recognize and clarify the rights of Indigenous Peoples.

Although not as dominant as they once were, fisheries are still a significant part of the formal (monetized) economy. In 2016, fish harvesting (including aquaculture) was a $4.7 billion industry, and fish processing generated another $6 billion. More than 75,000 Canadians are employed in fishing, aquaculture, or fish processing, about two-thirds of those in the four Atlantic provinces.

3.2 SALMON IN BRITISH COLUMBIA

Salmon play an important role in the ecosystems of the Pacific northwest (Map 3.1). Salmon are an anadromous species, meaning that they are born and spawn in a freshwater (upstream) environment but spend most of their lives growing in marine (ocean) environments. As such, they collectively act as "a conveyor belt for nutrients" (Post, 2008, para. 3), bringing nitrogen and phosphorous and other nutrients from the ocean upstream where they enrich terrestrial ecosystems. This movement adds a further complicating layer to attempts to govern or organize this resource.

Since time immemorial, Coast Salish Peoples have lived on the fish-rich coast of Vancouver Island and mainland British Columbia. Salmon were – and are – a central part of the subsistence, cultural, and economic fabric of the communities on that coast, and fishing techniques like dip netting in rivers and gillnetting and tidal traps at the mouths of creeks were used to harvest fish that were not only eaten immediately but also "smoked and dried, and later traded throughout large commercial networks that extended far beyond the immediate settlements" (Menzies & Butler, 2008, p. 138). If Indigenous relationships to salmon are, broadly, characterized by interconnectedness and relations, the European approach was decidedly more focused on ownership and extraction. Indeed, property regimes introduced and enforced by European settlers were based on the idea of bounded land privately owned by individuals, and settler regimes classified the living and non-living things within their boundaries as being owned by a specific individual (different conceptualizations and treatments of land are discussed in more detail in Chapter 7).

Such an arrangement facilitated resourcification of salmon. As Doug Harris (2006) documents, "The creation and imposition of a Native land policy that separated land reserved for Native peoples from that available to everyone else" (p. 51) was a colonial priority. These changes amounted to a colonial dispossession of land from Indigenous Peoples that fundamentally restructured the relationship between people and salmon on the Pacific coast. The result of settler colonialism in British Columbia was that Indigenous Peoples were displaced from the lands they had occupied for generations onto smaller coastal reserves, with the presumption that access to fish

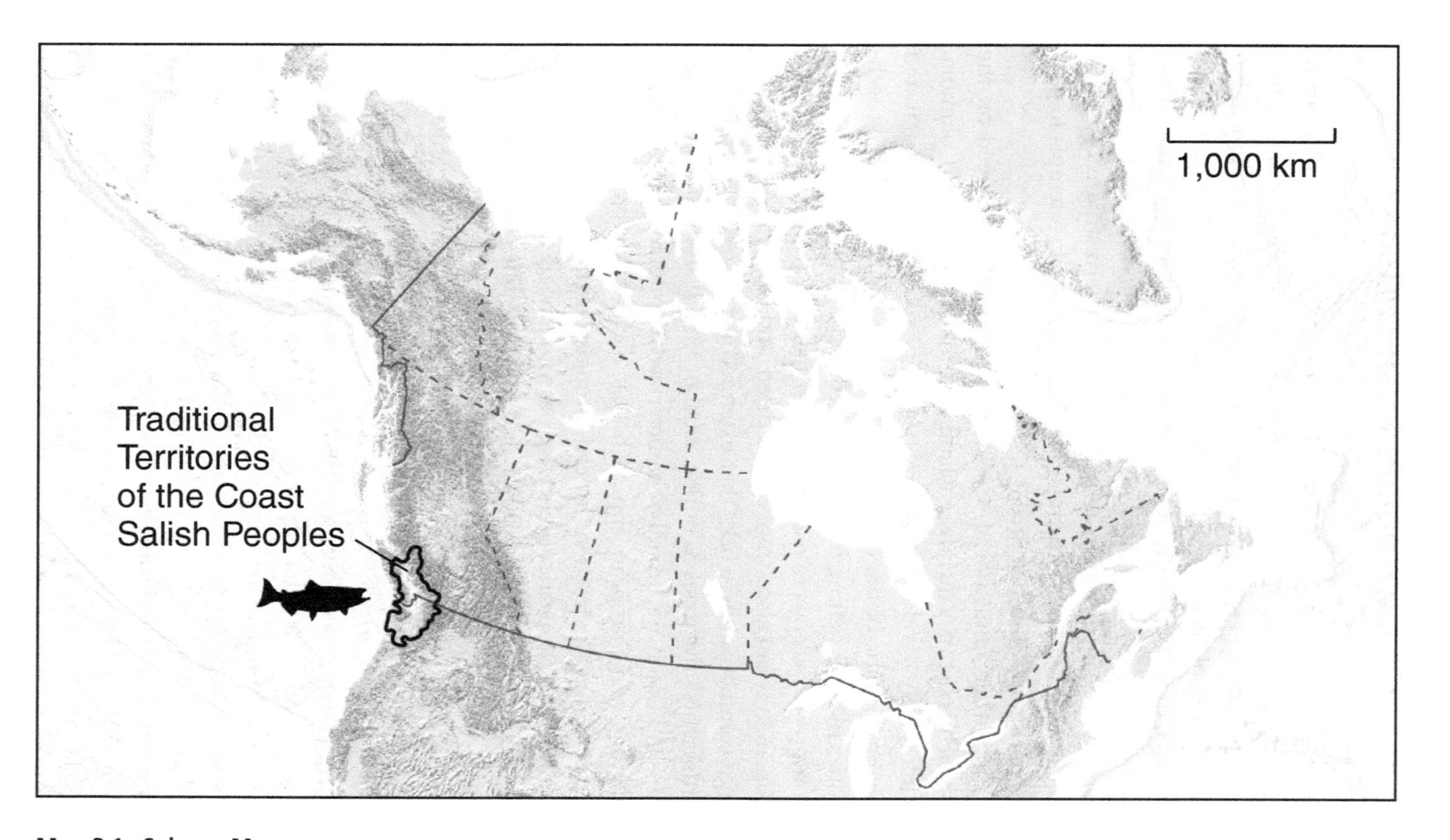

Map 3.1. Salmon Map

would continue – indeed, "half of the reserves allotted between 1850 and 1927 were specifically connected to fisheries .... Given the small land base, fish were the one resource that might have enabled Native peoples to build viable reserve-based economies" (Harris, 2006, pp. 51–2). However, the state then "imposed laws that limited Native peoples' access to the fisheries" (p. 52), namely by creating a separate Indian food fishery and leaving non-Native commercial interests to control the rest of the fisheries. Speaking to this process, Menzies and Butler (2008) write that "the 1888 Canadian Fisheries Act made a distinction between a registered Indian's right to fish for the purpose of food (which was exempt from certain regulations), and the right to sell, trade, or barter fish. This distinction was based upon erroneous mainstream colonial conventional wisdom that the selling of fish or the trade of fish for benefit was not an Indigenous practice. The effect of this regulation was to facilitate the incorporation of Indigenous fishers within the growing capitalist fishery" (p. 142). This integration took the form of hiring Indigenous labour in the burgeoning salmon cannery business, which was established, quite literally, on top of the traditional fisheries of the Indigenous Peoples of British Columbia's north coast (Menzies & Butler, 2008).

The first of 40 salmon cannery sites to be established over 8 decades was built on the Skeena River in 1876 and profited from seasonal Indigenous labour provided by families migrating to the canneries to take advantage of employment opportunities. An important note is that most of the fish in the canneries were fished by knowledgeable and experienced Indigenous Peoples hired to captain the settlers' boats on the settlers' licences. This arrangement further entrenched the divides between settler and Indigenous populations when a 1912 Memorandum of Understanding (MOU) between British Columbia and the federal government was drafted with the goal of removing Indigenous fishers from the commercial fishery, stating that "it is eminently desirable to have the fisheries carried on by a suitable class of white fishermen" (Menzies, 2016, p. 103). As Menzies and Butler (2008) note, these efforts "worked to exclude indigenous peoples as full participants in the market economy as owners while relegating them to sources of labour power to be extracted analogously to the natural resources of fish, trees, and minerals" (pp. 142–3).

To put this MOU in context, it came at a time when many commercial fishers were, in fact, Indigenous. As Wright (2008, p. 100) notes, the fishing industry was the largest single employer of Indigenous Peoples in the province for much of the twentieth century. As the MOU painfully shows, this was something with which many settlers took issue, and, through mechanisms both legal and financial, Indigenous Peoples were dispossessed of a fishery they had managed for generations. This squeeze occurred as the non-Indigenous fishery became more capital intensive, with settler fishers having easier access to financial credit and larger boats. This more capital-intensive fishing was bolstered by state regulation that tightened the rules about when, where, and how people could fish.

These processes of dispossession also connect to this book's theme of the nature–society binary and Canadian imaginaries. Specifically, the 1888 segregation of the Indigenous food fishery from other fisheries "served to separate Indigenous communities from the wealth of their salmon resources while conserving them for the non-Native commercial fishery" (Foster, n.d., "Not a Race-Based Fishery" section). When Indigenous people working for wages in the industrial fishery integrated this seasonal employment into mixed cash–subsistence economies, it did not pay for the sale of the resources, nor did Indigenous communities ever give up ownership or management control of their salmon runs (Foster, n.d.). Here, in a creative bit of colonial thinking, commercial fisheries are envisioned as part of the economy, and Indigenous fisheries are envisioned as something separate from the economy. Drawing these boundaries – between commercial and subsistence fishing and between commercial and Indigenous fisheries – serves the ends of colonialism and economic growth to the detriment of Indigenous Peoples and salmon.

If the salmon fishery of the late 1800s and early 1900s was characterized by settler hostility toward Indigenous fishing communities, the salmon fishery in the latter half of the 1900s was characterized by tensions arising from a previous colonial activity: the drawing of the Canada–US border. Indeed, disputes over Canadian and American fishing rights were at the heart of the signing of the 1985 Pacific Salmon Treaty and a series of disputes in the early 1990s about each country's allowable catch in the waters shared among British Columbia, Washington State, and Alaska. Perhaps most notably, the 1994,

1995, and 2004 salmon runs were significantly lower than predicted, leading to fingers being pointed at all involved, with the Indigenous food fishery (IFF), scientists, and US politicians and boats taking most of the blame.

In 2009, disaster struck when only 1.4 million sockeye salmon returned to the Fraser River; more than 11.4 million had been expected (B. Cohen, 2012, p. 120). In response to this decimation, Prime Minister Stephen Harper appointed BC Supreme Court Justice Bruce Cohen to chair a commission to look into the missing salmon. The enquiry held several months of hearings, collected more than 3 million pages of documents, heard from 179 witnesses, and cost $25 million (CBC News, 2012). This exhaustive study determined that the sudden crash in salmon population could not be blamed on a single factor. Rather, a suite of (mostly) human-caused changes to ocean conditions had worked together to make the salmon's habitat less hospitable. These stressors included harmful algal blooms, low food availability, freshwater and marine pathogens, disease, and warming ocean temperatures resulting from climate change. In particular, Cohen (2012) spoke to the potential effects of salmon farms lining the West Coast, acknowledging the potential role that sea lice and waste from the farms may have played in wild salmon populations and also noting that more research is needed. Salmon runs since the 2009 crisis have varied year to year, and more salmon farms have come online as market demand for salmon remains strong despite variable wild catches.

The commission's report is relevant to the arguments presented in this book because it points to a collection of concrete ways in which humans have altered their environment and, more important, how those alterations are driving environmental changes that will force humans to reorganize themselves, thereby exemplifying the co-constitutive relationship between people and place that we highlight throughout this book. Specifically, a precipitous decline in BC salmon populations would be – and indeed, is – part of a reorientation of British Columbia's export-focused, resource-based economy away from wild fish and forestry and toward fish farming and energy development (G.N. Wilson & Bowles, 2015). Moreover, the case of salmon in British Columbia highlights an important instance of binary creation relevant to our analysis in this book: the creation in 1888 of a legal distinction between the IFF and the commercial fishery. The creation of two sets of legal rights to the same fish has been responsible for significant tensions between Indigenous and settler communities (D. Brown, 2005); it also creates and reinforces the idea that somehow fish stocks can be divvied up and managed separately. In its racist inception, it may have mirrored the nature–society binary inasmuch as the Indigenous fishery was imagined to be natural, but today the distinction and implementation of the IFF – which certainly echoes its racist past – is a reflection of treaty rights. Although this distinction was framed in 1998 by a protesting group of non-Native fishermen as a race-based fishery, the Supreme Court of Canada rejected this argument (Foster, n.d.) because access to the IFF is accorded only on the basis of individuals' treaty rights. In this sense, it is a rights-based fishery. Nevertheless, the creation of the IFF is an example of the kinds of binary creation that are the focal point of this book.

3.3 COD IN NEWFOUNDLAND AND LABRADOR

In a Heritage Minute video that dramatizes John Cabot's first encounter with abundant cod in the late fifteenth century, he exclaims to his patron (King Henry VII) that it would provide "fishes enough to feed this kingdom ... until the end of time!" (Historica Canada, 1991a; see the Pedagogical Resources at the end of this chapter for links). Although it did not last until the end of time, the Atlantic cod fishery (Map 3.2) did prove to be an enormous boon to Europeans: a resource that provided wealth and caloric fuel for imperial ambitions over generations.

Early in the development of European cod fishing, communities on the Atlantic coast of North America were established, especially on the island of Newfoundland: ports that fishing ships would return to, where fishers could preserve their catch by drying it. Although some continental European fishers had access to abundant salt that allowed them to preserve their fish onboard ships, English fishers, for whom salt was scarcer, relied on a land base proximate to the fishing grounds to get their fish to Europe in edible form (Ryan, 1990). One government report summarizes the history by saying that "northern cod was the *raison d'être* for the existence of Newfoundland as a colony and later as a Dominion" (Emery, 1992, p. 1). Cod, along with

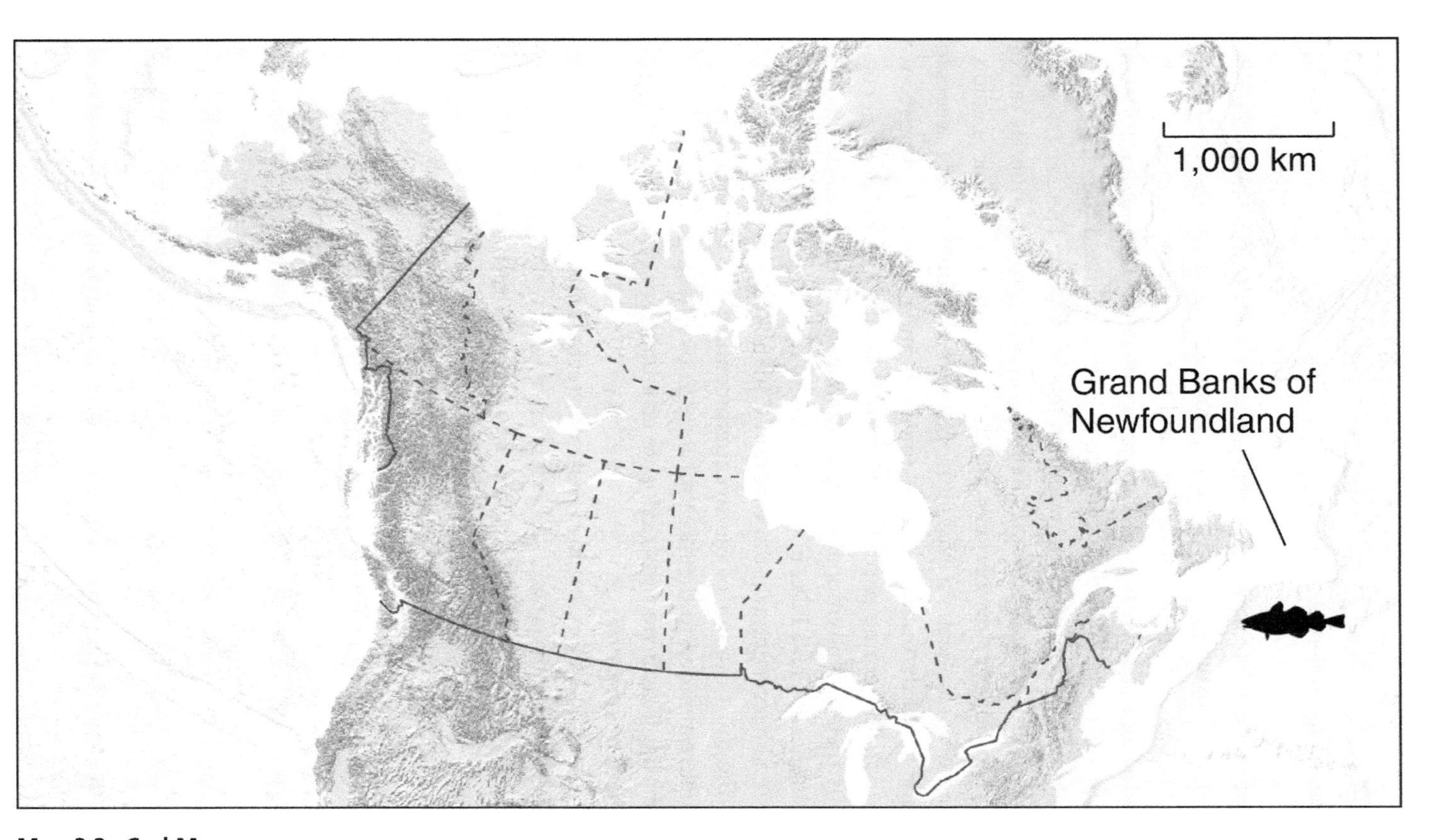

Map 3.2. Cod Map

ships, fishing lines and hooks, and salt, among other things, was a resource that organized Newfoundland outport communities for more than half a millennium.

As exploitation of the fishery intensified, it evolved from being labour intensive to being capital intensive. Over the nineteenth century, dominant fishing practices shifted from sailing or rowing vessels using hooks and lines, to longlines, and then to steam-engined trawlers that dragged large nets along the ocean floor. Technological developments in the twentieth century in turn globalized the cod fishery, making local outport communities less necessary. Freezer trawlers, able to process and preserve (freeze) fish on board, can operate out of port for longer time periods and thus over greater distances, allowing European fleets to exploit the resource without needing access to ports in North America. At the same time, and particularly since Newfoundland became part of Canada in 1949, the cod fishery has been constructed as a national resource managed by the Canadian federal government, and FOC in particular. Thus, as a greater proportion of the cod harvesting was done by non-Canadian vessels over the course of the third quarter of the twentieth century, the Canadian government responded to the threat of foreign overfishing by extending its fisheries jurisdiction (its exclusive economic zone) to 200 miles (322 kilometres) in 1977.

As with salmon, the idea that cod stocks could be divided among different countries and separately managed was implausible. Codfish continued to live their lives in ignorance of these jurisdictional lines, which, for humans, are more difficult to see and police than land borders. Since 1979, fishing outside the 200-mile limit has been managed by an international regulatory agency, the North Atlantic Fisheries Organization (NAFO; Emery, 1992, pp. 7–8). The NAFO regime has, however, not been an effective one, and cod overfishing remained a persistent problem. The ineffectiveness of the international management regime led to the 1995 "Turbot War," in which Canadian vessels responded to concerns about foreign overfishing just outside of the Canadian 200-mile limit, forcibly detaining a Spanish trawler in international waters for using a net that was illegal in Canada. Although the Spanish ship was operating in international waters using a net that was legal under European Union regulations, the Canadian government's actions in this case clearly suggested that management of the cod was seen as a Canadian responsibility, even outside the 200-mile limit.

The Canadian government's defence of fish outside of Canadian territorial waters could be seen through a conservationist lens: protecting a declining species in the face of rapacious extractors who happen to be from another country (Springer, 1997). Certainly the federal government was not above presenting itself in that light, as then-Minister of Fisheries Brian Tobin declared, "We're down now finally to one last, lonely, unloved, unattractive little turbot clinging on by its fingernails to the Grand Banks of Newfoundland" (DeMont, 1995, as cited in Springer, 1997, p. 47; for a non-Canadian view of the dispute, see Swardson, 1995).

However, the extension of Canadian jurisdiction was not only intended to conserve fish stocks; the 1977 move to assert jurisdiction up to 200 miles from the shore was also seen as an opportunity for Canadian industry growth, because it was "accompanied by a wave of optimism and highly leveraged capital investment in both fishing vessels and processing plants" (Emery, 1992, p. 10). By the early 1990s the fishery "produced 20 percent of the GDP of the Atlantic region, employed 100,000 people, and supported 1,500 communities" (D. Brown, 2005, p. 50). The value of fisheries production across the country exceeded $3 billion in 2016 (DFO, 2018).

The extension of the reach of Canadian law (jurisdiction) to 200 miles and beyond, including risking an international incident by detaining a Spanish ship in international waters, are examples of nation building in a literal sense: pushing out the country's borders to encompass previously unclaimed territory. It is also an exercise in nation building in a more symbolic sense, reinforcing a sense of national identity or solidarity by focusing on the economic development of Canada's resources or by focusing ecological conservation on a shared enemy (foreign trawlers) rather than regulating domestic producers.

As well as increasing the extensive reach of the Canadian state outward, the history of cod fishing also shows the increasingly intensive reach of the Canadian state inward. Over the second half of the twentieth century, the Canadian government engaged in increasingly detailed fishery regulation and management, intended to ensure the sustainability of the fishery resource and to maintain fishers' incomes. Although there were conservationist elements to these efforts, it is

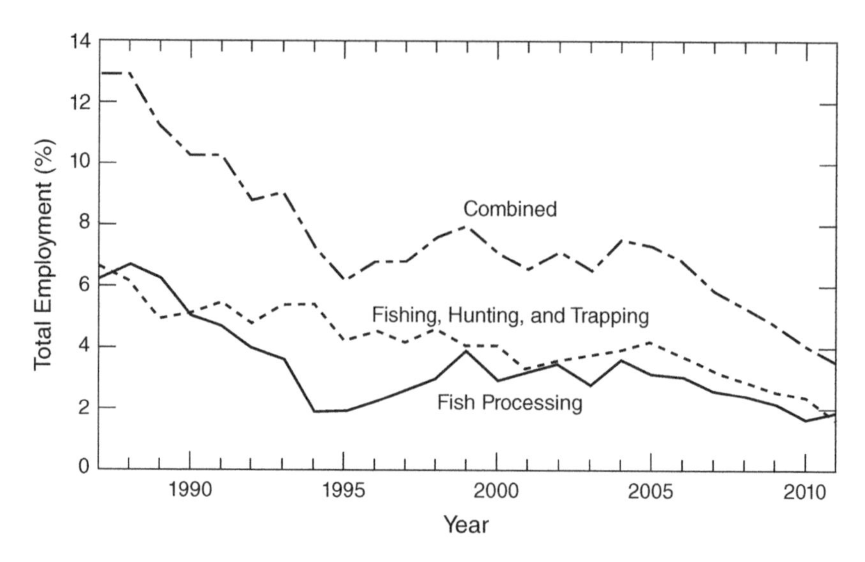

Figure 3.1. Newfoundland and Labrador Employment Chart

important to note that the conservationism remained embedded in resource thinking. Scientists and managers did not see fish as components of an ocean ecosystem so much as they saw a fishery for human (and more specifically Canadian) use. Moreover, governments assumed that fish populations could be sustained at the same time as the fishing industry could continue to grow. Perhaps not surprisingly, these efforts to produce sustainable growth ultimately failed. By the 1990s, scientists estimated that biomass of northern cod had declined by an incredible 99 per cent (Myers et al., 1997). Even these studies of environmental collapse are structured by resource thinking in their terminology and ideas, using terms such as *biomass*, which refers to the estimated volume of fish aged three years and older, and *northern cod*, which refers to cod off the coasts of Labrador and northeastern Newfoundland. In response to this decline, a moratorium on cod fishing was imposed by the federal government in 1992 (Figure 3.1). The moratorium put 35,000 Newfoundlanders out of work, an incident described as "one of the world's greatest fisheries management failures" (Bavington, 2010, p. 523). "A Great Destruction" (1996, para. 6) notes that, at the time, one-sixth of Newfoundland's workforce "depend[ed] on the fishery for some or all of its income" and that "if a calamity of similar magnitude befell Ontario's manufacturing industries, some 800,000 people would lose their jobs."

The collapse of the cod fishery itself, which precipitated the moratorium and layoffs, "could be Canada's greatest environmental failure to date" (Olive, 2019, p. 16). To an even greater extent than the salmon collapse on the West Coast, discussed in the preceding section, the cod collapse forced a major restructuring of the Newfoundland economy, as well as the various other channels through which resources were organized, including government institutions, built environments, and culture. Because cod was no longer (or to a significantly lesser extent) available as an organizing resource, new resources had to be found. Along with the rise of even more intensively managed aquaculture operations, such as fish farming (Bavington, 2010, pp. 523–4), there has been a shift to offshore oil and gas extraction (including the offshoring of workers from Newfoundland to the Alberta oil sands). Perhaps most notable has been the rise of tourism, insofar as cod has been reinvented as an iconic resource in the "screeching in" ceremony that attracts tourists eager to experience not only Newfoundland's wild nature but also the organic culture of authentic small fishing communities.

For our purposes, two broad lessons can be drawn from the collapse of the cod fishery. The first is that, in the inevitable tension between sustainability and growth, the process of fish resourcification prioritizes growth at the expense of sustainability. As fish resources were organized more effectively, the fishery grew. That this organization of the fishery was effective, to a point, is shown in the data about the size of the late twentieth-century fishery, mentioned earlier. This success also produced political actors with an interest not only in sustaining fish but also in increasing the revenue to be derived from the fishery. The management of renewable resources always has to contend with ensuring that the rate at which a resource is harvested does not outstrip the rate at which it is being regenerated. However, as decisions about resource extraction are increasingly made in the context of institutions that are impersonal and not locally embedded, the pressure for continuous growth becomes increasingly difficult to resist. (We explain this point in more detail later.) This is true when those non-locally embedded institutions are globalized markets (including the market for fish themselves, but also the market for investment capital; for boats, licences, processing plants; and so on; Markham, 1994). It also applies to state institutions. For example, the 1990 report of the federal government's Northern Cod Review

Panel discussed how political calculations to appease various interests can lock in unsustainable trajectories of resource consumption: "The temptations to grant the licences or to approve the loans may be nearly irresistible. But, so may be the pressures subsequently generated to allocate the resources to justify the earlier decisions. The repercussions may be disastrous for the stocks" (Harris Report, 1990, quoted in Emery, 1992, p. 15).

A second, related lesson is that resource thinking can drive technological innovation, which can in turn exacerbate unsustainability. The increasing technological sophistication of fishing practices, along with the increasingly detailed regulation and management of fishing as an industry, were tied in with the deepening conceptual shift from fish to fishery, or from ecosystem component to resource. Over the course of the first half of the twentieth century, biologists increasingly sought to understand fish through the lens of population rather than species. As Dean Bavington (2010) describes it, this represented a "switch from qualitative typological species descriptions of fish (observed through the senses or based on experiences out on the water) to statistical population analysis (mathematical constructs accessible only to experts)" (p. 511). In other words, there was an increasing sense that the best kinds of knowledge about fish came not from fishers or those who observed and interacted with them directly, but from those with the capacity to study them from afar, who, because they were not embedded in local environments, could see the big picture. Fishing had previously been understood as a practice that was subject to inevitable natural variation, with catches fluctuating quite widely from one year to the next, and a resultant boom–bust economic cycle. Population analysis was seen as a tool to solve this problem. As noted earlier, these abstract models produced wondrous results – at least at first. As fishing methods became increasingly intensive, and fishing economies became increasingly competitive, by the mid-twentieth century scientific population management was being used to calculate maximum sustainable yields (MSYs; pp. 513–4). In terms of the Atlantic cod fishery, Bavington puts it starkly: "For the first time in history, cod appeared to be manageable objects ... whose productivity could be influenced by carefully using fishing to kill specified numbers of surplus fish in separate, identifiable populations" (pp. 514–5). To produce this abstract model, however, "a staggering amount of knowledge about wild codfish had to be omitted to achieve this manageable representation" (p. 517). At the same time, the regulation and organization of the fishing industry – the work of human fishers – was similarly remade in the light of abstract models, in this case developed by economists. "By the 1980s ... fish and fishermen had been conceptually domesticated, represented as quantifiable elements in predictable and controllable bio-economic models" (p. 521).

As noted earlier, however, the resource was nonetheless overfished, and despite scientific management and government regulation that seemed to be focused on sustainability, the cod population collapsed disastrously. Or perhaps this occurred not only despite scientific management and government regulation but also because of it. The use of scientific models and impersonal regulation, through government bureaucracy, global markets, or both, means that locally embedded knowledge is abstracted away and cannot be seen. In the case of the cod fishery, inshore fishers, who had always needed greater local knowledge of when and where fish could be found, had sensed that fishing levels were unsustainable, but their warnings went unheeded (Berkes et al., 2000, p. 1257). In short, the material and conceptual organization of fishing resources that facilitated the growth of the industry also had a blind spot. The apparent paradox is that organizing nature more intensively and scientifically may be just what drives resource use to unsustainable levels.

3.4 CHANNELS IN ACTION: ORGANIZING FISHERIES

Fish (ecosystem components) have been turned into fisheries (resources) and, once organized, those resources have been used to further reorganize socioecological systems in Canada. This organizing happens through several different channels, including but not limited to the formal politics that happen in government institutions.

The institutions of **government** provide an important channel for this transformation. The institutional scaffolding of fish governance in Canada can help explain how ecosystem components have been transformed into capital and the consequences of this transition. Although individual pieces of legislation and policy decisions can be important and highly visible vectors for making political change, our focus on government here is mainly on the administrative structure of the Canadian state, as well as the role of the courts. How the state itself is organized is important because this organization sets conditions and limits on the kinds of individual decisions that can be made, favouring some actors and kinds of decisions over others.

One important way in which institutions have played a role in this transformation is in the arrangement of the institutions themselves, which promotes the nature–society binary within federal departments. As many scholars have noted, the Canadian Constitution is notoriously ambiguous when it comes to environmental responsibility: the word *environment* is completely absent from the Constitution, such that many of today's environmental concerns are exacerbated by a lack of clear governmental responsibility (see, e.g., Harrison, 1996; Paehlke, 2004; Weibust, 2009). Fisheries, however, are a notable exception to this rule. Under Section 91 of the Canadian Constitution (*Constitution Act*, 1867), the federal government is responsible for sea coast and inland fisheries. Indeed, founded on July 1, 1867, the Department of Marine and Fisheries (now FOC) was one of Canada's first federal departments.

Although fisheries are clearly an area of federal jurisdiction, there is shared responsibility within the federal government. Specifically, Environment and Climate Change Canada (ECCC) is responsible for administering Section 36 of the *Fisheries Act*, which protects fish habitat from pollution. (Other habitat concerns, such as dams or other development on fish-bearing waterways, are the responsibility of FOC.) Section 36 prohibits the depositing of deleterious (harmful) substances in water "frequented by fish" (*Fisheries Act*, 1985, s. 36[3]). To that end, ECCC is responsible for administering specific regulations addressing the types, concentrations, frequencies, and locations of effluent allowed in fish-bearing waterways. For example, it regulates effluent from meat and poultry product production, from mining operations, from petroleum refining, from pulp and paper processing, and from municipal wastewater systems. So, fisheries are clearly a federal responsibility, and their administration under the *Fisheries Act* is shared between FOC and ECCC.

This division of responsibilities is emblematic of the kinds of nature–society divides that are critical to the argument we make in this book. To understand why, it is useful to look at the official mandates of FOC and ECCC.

FOC fulfills its mandate by

- *sustainably managing fisheries and aquaculture*
- *working with fishers, coastal and Indigenous communities to enable their continued prosperity from fish and seafood*
- ensuring that Canada's oceans and other aquatic ecosystems are protected from negative impacts
- ensuring commercial vessels and recreational boaters can safely navigate our waters (DFO 2022, emphasis added)

ECCC's mandate, however, extends to matters such as

- *the preservation and enhancement of the quality of the natural environment, including water, air and soil quality, and the coordination of the relevant policies and programs of the Government of Canada;*
- renewable resources, including migratory birds and other non-domestic flora and fauna;
- meteorology; and
- the enforcement of rules and regulations. (ECCC, 2021, emphasis added)

The italicized text is telling: whereas FOC's mandate is to manage fisheries (i.e., society), ECCC's mandate is to protect the natural environment (i.e., nature). This separation tells us something about the transition from ecosystem component to commodity: once a fish is caught, it is no longer considered part of the environment but is instead part of a fishery. Indeed, as Bavington (2010) notes in the case of Newfoundland cod:

> Before the advent of fisheries management, cod were understood as a free species embedded in an uncommodified common ocean – nobody owned codfish until they were hunted down and pulled onboard a fishing vessel. After the spread of population thinking, the development of bio-economic models, and the

> establishment of the 200-mile [exclusive economic zone], *cod became members of large swimming inventories* whose current and future number could be assessed onshore and allocated as biomass to fishing operators long before they were actually killed by a fisherman or they had spawned in the actual ocean. (p. 522, emphasis added)

What this suggests is that as resource thinking expands and resources become more highly organized, fish become part of the fishery, not part of the environment, before they are caught or even before they are born. Here one can see the influence of **culture and ideas** – in this case, scientific knowledge – in organizing fish as resources.

In addition to reflecting and deepening a false nature–society binary, the reconceptualization of fish from ecosystem components to economic resources worked in favour of institutionalized nation-building efforts: fishing generated income for individuals, communities, and governments and bolstered sovereignty claims by spurring the creation of exclusive economic zones 200 miles from the Canadian coastline and focusing on foreign ships as the cause of the cod's decline. More important, the creation of the IFF in 1888, as detailed in Section 3.2, contributed to the false nature–society binary, too, by separating commercial fishing from all other activity.

The centrality of fishing to many Indigenous **identities** is increasingly being recognized through the Canadian court system. Fishing rights are often among the rights described in the various treaties that Indigenous nations made with the Crown and are constitutionally protected. It is the job of Canadian courts to interpret existing legislation (Box 3.1). Doing so creates legal precedent, which clarifies the rights and responsibilities of various parties as they relate to a particular piece of legislation. For this chapter, it is important to note that many of these landmark cases about Indigenous identities and Aboriginal rights have involved questions about fishing, and the courts' rulings have applied not only to fishing but to a broader suite of issues (see Table 3.1).

At least to the extent that Aboriginal fisheries are oriented around the pursuit of a moderate livelihood (or personal and ceremonial uses) rather than the pursuit of (in principle, unlimited) growth and

Table 3.1. Selected Fishing-Related Supreme Court of Canada Cases

Case	Year	Importance
Calder v. Attorney-General of British Columbia	1973	The "first time that the Canadian legal system acknowledged the existence of Aboriginal title to land and that such title existed outside of, and was not simply derived from, colonial law" (Salomons, n.d., para. 2).
R. v. Sparrow	1990	"Set out criteria to determine whether governmental infringement on Aboriginal rights was justifiable, providing that these rights were in existence at the time of the *Constitution Act* (1982). These criteria are known as the Sparrow Test" (Salomons & Hanson, n.d., para. 1).
R. v. Van der Peet	1996	Further defined Aboriginal rights per Section 35 of the Canadian Constitution as rights that are "integral to the culture of the claimant" (Hanson & Salomons, n.d., para. 3). The Van der Peet test, or the Integral to a Distinctive Culture Test, outlines 10 criteria required for a practice to be protected as an Aboriginal right under Section 35 (Hanson & Salomons, n.d., para. 3).
R. v. Marshall	1999	Recognized, on the basis of the Peace and Friendship Treaties of the mid-eighteenth century, the rights of the Mi'kmaq, Wolastoqiyik, and Peskotomuhkati Peoples to commercial and personal sustenance fishing. It is important to note, however, that this right to commercial fishing was limited to a "moderate livelihood" and not "the open-ended accumulation of wealth," which has meant that Indigenous fishers have remained at least to some extent excluded from the larger-scale, corporatized, and profit-oriented commercial fisheries (Blackmore, 2019).

accumulation of profits, they may be better positioned to resist some of the problems of resource thinking. The emphasis on moderation, or what Thomas Princen (2005), in a non-Indigenous context, calls "sufficiency,'" is distinct from, but related to, an Indigenous cosmological perspective that emphasizes the relatedness or connectedness of human and non-human beings.

BIG IDEAS IN SMALL BOXES

BOX 3.1. THE JUDICIAL BRANCH OF CANADIAN GOVERNMENT

In Canada, the legislative branch of government (i.e., the House of Commons and Senate, the provincial and territorial legislatures) creates laws, and the judicial branch of government (i.e., the courts) interprets them. In other words, it is the job of the courts to determine when a law has been broken. This may sound straightforward – and in many cases it is – but Canada's laws and statutes often leave room for interpretation. For example, Section 35 of the Canadian Constitution (*Constitution Act*, 1982) states that "the existing aboriginal and treaty rights of the aboriginal people in Canada are hereby recognized and affirmed." This one sentence alone raises many questions. For example, which rights are existing, and what does it mean to recognize and affirm them?

In Canada, these kinds of questions are addressed when specific legal questions make their way into the court system. That is, the court will not address these questions in a vague, abstract way but will interpret legislative text in the context of an actual legal dispute. In Canada's common-law system, case law is used to resolve disputes, and the courts' judgments are constrained by (1) existing legislation (i.e., courts can only rule on the violation of existing legislation, not on something that should exist), (2) previous legal decisions (i.e., a court ruling must be consistent with past rulings on similar cases), and (3) congruence with higher courts (e.g., a provincial court's rulings must not violate previous Supreme Court case decisions). Because precedent is so central to the Canadian legal system, a single case can have an impact that goes well beyond the individuals involved in that particular case. For example, in *Halpern v. Canada (Attorney General)* (2002), Ontario's Court of Appeal in 2003 ruled that the legal definition of marriage as being between one man and one woman violated Section 15 of the *Canadian Charter of Rights and Freedoms* (1982), which guarantees equality rights. As a result, same-sex marriage became legal in Ontario. Other provinces followed suit, and in July 2005, the federal government enacted the *Civil Marriage Act* (2005), which extends the legal capacity for marriage to same-sex couples across the country and allowed thousands of same-sex couples across the country to marry.

In a similar way, a small number of landmark cases have established the legal frameworks for Indigenous rights in Canada. Broadly, the past 40 years has seen a trend of the Canadian courts interpreting legislation in ways that strengthen and sharpen Indigenous rights. These important advancements are discussed in the relevant case study chapters.

Of course, the institutional structures of government and decisions by formal political actors like judges or politicians are not the only things that shape how fish(eries) are turned into resources and governed. **Economies**, in which money rather than political office is the medium of power, are also powerful channels for turning ecosystem components into resources.

As the two examples discussed in the first half of this chapter suggest, the unsustainable growth and collapse of resources happens for multiple reasons that are often interrelated in complex ways. The specific trajectory of overdevelopment followed by exhaustion is at least to some extent unique for each particular resource, although in both cases, the economy is an important channel for the flow of organizing power. Cod and salmon are different species, with different migration patterns and different vulnerabilities to disease, environmental change, and other threats. How stocks of the different species could be governed were thus different, as were the technologies used to capture and process them into food resources. Indigenous–settler relations developed differently on the Atlantic coast, the first point of contact, than on the Pacific coast. Foreign threats on the Pacific coast were limited to US fishers, whereas on the Atlantic coast a wider range of countries was involved. Although they were unable to prevent the collapse of salmon stocks in the late 2000s, some of those involved in governing the West Coast fisheries had nonetheless learned lessons from, and adapted practices in light of, the cod collapse on the East Coast in the 1990s.

Despite these differences (and many more could be mentioned), some general features and trends can be discerned. Both the cod and salmon fisheries are exemplary of a boom–bust cycle that often characterizes resource economies, in which extraction grows to unsustainable levels and then collapses. It is perhaps easier to intuit why this happens with non-renewable resources, such as minerals: if there is a finite stock available, then no level of extraction is indefinitely sustainable. A boom in the extraction of any non-renewable resource inevitably leads to a bust: at some point, it will no longer make sense to extract the resource in question, perhaps because it runs out, or market prices drop, or technology changes, or consumer demand evolves. (In fact, most busts occur because of changes in the market for a resource, rather than an exhaustion of the physical supply.) Why, though, does this boom–bust cycle also seem to characterize renewable resources such as fish that, at least in theory, could reproduce themselves indefinitely?

Garrett Hardin (1968) suggested that this phenomenon, which he called the "tragedy of the commons," was an inevitable feature of resources to which access was not rigorously controlled. Presuming that humans were naturally and relentlessly appetitive and individualistic, Hardin thought people would overexploit resources whenever they could, because the individual gain from short-term overexploitation would outweigh the collective benefits of long-term restraint. Hardin's idealized example is a commonly owned grazing pasture: each farmer sees the advantage of putting additional cattle onto the pasture, to the point at which overgrazing leads to ecological collapse. For Hardin, the only way out of this predicament is "mutual coercion, mutually agreed upon" (p. 1247). In practice, this has usually been interpreted (including by Hardin himself) as prescribing either authoritarian regulation or the privatization of the resource, so that it is individually rather than collectively managed.

It is important to note, however, that many **communities** have sustainably managed common-pool resources (such as fisheries) over very long periods of time. Elinor Ostrom (1990) showed that there are many examples of communities that more or less sustainably managed commonly held resources without top-down regulation. Drawing on these examples, Ostrom identified eight design principles that needed to be present to achieve this (Box 3.2). The regime for managing Canadian fisheries at least to some extent follows Hardin's (1968) prescription ("mutual coercion, mutually agreed upon"): a quota (total allowable catch) is established and enforced by a single central authority (FOC). In doing so, it is at odds with at least some of the principles identified by Ostrom, such as the ability for local fishers to have a say in making the rules (principle 3) and having their local autonomy respected (principle 4). Hardin and Ostrom thus provide two very different sets of ideas or principles for managing resources in a sustainable way: ideas that are taken up by various actors in the struggles over how to organize fisheries.

Whereas Hardin (1968) suggests that the roots of the boom–bust cycle are (what he presumes is) an innate feature of human nature, the many examples of community management that Ostrom (1990) identifies suggest otherwise. Instead of blaming human nature in general, one might instead look for those roots more specifically in the organization of the economic system, which is predicated on always looking for booms. Canada's economic system (capitalism) is one in which most decisions over the allocation of productive resources (fishing boats, trawlers, processing plants, etc.) are in private hands. It is private actors – individuals or corporations – who decide whether or when to put boats and nets in the water. Of course, they are subject to important forms of government regulation and oversight, particularly in the form of catch quotas and a limited number of fishing licences. Nevertheless, government only provides the opportunity to deploy those resources: literally, a licence to fish. What gets someone to actually use that licence, or open a processing plant, is the possibility of realizing a financial gain by doing so. Particularly for those who are in communities dependent on a single resource, this is often not really much of a choice: work in the fishery may be the only viable option in that community to sustain oneself and one's family. In this context, chasing the next boom becomes a structural feature of such an economic system – for individuals and for companies – and failing to be the next boom means bust.

Boom-and-bust economies are vulnerable, as the cod collapse demonstrates, and they are an inevitable feature of an economy that is unplanned and driven by a quest for financial gain that is in principle unlimited. It is important to note, however, that from a certain perspective, this boom–bust dynamic is a feature rather than a bug. It is what makes capitalist economies dynamic and enforces a relentless pursuit of increased productivity. It is, in other words, what drives growth in the economy overall. In this sense, capitalist economies are not only prone to periodic crises, they are also dependent on them (Fraser & Jaeggi, 2018; J. O'Connor, 1973).

BIG IDEAS IN SMALL BOXES

BOX 3.2. DESIGN PRINCIPLES FOR COMMON POOL RESOURCES

Elinor Ostrom (1933–2012) was an economist who studied the diverse institutional arrangements that communities used to govern common pool resources – resources that are commonly owned and used. Rather than deducing proper institutional arrangements from abstract theoretical principles or assumptions, Ostrom started by examining existing institutional arrangements around the world and with a variety of resources, and from there she induced a set of eight design principles that characterize sustainable resource management. For this research, Ostrom won the Nobel Prize in Economics in 2009, the first woman to receive this award. The eight principles are as follows:

1. Boundaries are clearly defined.
2. Rules are adapted to local needs and conditions.
3. Those affected have a say in making the rules.
4. Users' rights to make rules are respected by outside authorities.
5. Users' behaviour is monitored.
6. Rule violations are punished with graduated sanctions (i.e., harsher punishments for repeat offenders).
7. Conflict resolution mechanisms are easily accessible.
8. Where more than one level of governance is necessary, authority should be nested.

Built environments are in some ways a material marker of resource extraction. In cases of the boom–bust cycle, they may provide a physical trace that lasts long after resource extraction has ended. As noted earlier, Newfoundland fishing **communities** were among the very first settlements established by Europeans in North America. English fishers originally settled around the island to preserve their fish before returning to Europe. Fishing communities were built up on the Pacific coast, too, first by Indigenous Peoples and then transformed by settler colonialism. At a finer-grained resolution, the built environment was reshaped as the fishing industry became more capital intensive over time and as small-scale individual enterprises were replaced by larger-scale, more industrial operations, eventually culminating in a "large-scale, permanent, capital-intensive fishery" (Hoogensen, 2007, p. 47). As Hoogensen (2007) describes, "Modern factories (canneries) using indigenous and imported labour began to flourish on both the Atlantic and the Pacific coasts and on the Great Lakes and the large freshwater lakes in Canada's interior" (p. 46). As fisheries have scaled up in the Anthropocene era (see Box 2.2), human impacts on the marine environment have become increasingly pronounced. In the case of the salmon fishery, the Cohen Inquiry (2012) found human-caused changes to the ocean environment, with impacts ranging from local (fish farms) to global (climate change).

Reflecting the **culture and ideas** channel, these built environments also developed distinctive cultural environments, including distinctive forms of knowledge. As previously noted, some of the first European settlements in North America were organized around the extraction and processing of locally specific fish resources. By living and working in these communities that were built by and for fishing, fishers developed local geographic knowledge – understanding of cod migration patterns – that allowed them to successfully exploit the inshore fishery.

The ability to exploit any resource clearly depends on knowing where (and for mobile beings such as fish, when) it can be found. Such knowledge is developed and held by the people involved in resource extraction, a community that is often highly localized. In some cases, this knowledge is complex and developed over long periods of time. As Ryan (1990) notes in the case of the inshore cod fishery, "a knowledge of the fishing grounds took years to acquire, and was added to in each generation" (para. 4). Knowledge of salmon migration patterns held by Indigenous communities on the Pacific coast is similarly highly developed and held within particular local communities.

At least in these kinds of cases, where resource exploitation depends on extensive and slowly acquired local knowledge, resource communities are deeply rooted to a particular place: a counterpoint to the abstract scientific knowledges discussed earlier. Berkes et al. (2000) note that such traditional ecological knowledge (TEK) is "a cumulative body of knowledge, practice, and belief": knowledge is not just abstract or disinterested information but is intimately connected with "a component of practice in the way people carry out their resource use activities, and further, a component of belief regarding how people fit into or relate to ecosystems" (p. 1252). Because it is connected to practices and beliefs, this is not the kind of knowledge that can be gleaned simply by reading a textbook or watching an instructional video. To know in this context means to perform certain actions and to believe certain things about the world. In other words, the acquisition of knowledge involves immersion (to a greater or lesser extent) in a particular community. TEK itself is a resource that is held, or organized, at the community level, although, as the case of cod demonstrates, the value of that knowledge is not always recognized beyond that community. As we have shown, the shift to a more capital-intensive scientific management of fisheries as resources (populations) has been at odds with TEK that is locally embedded and developed through practical engagement with the environment.

As a key component of the diet since time immemorial, the rituals and traditions around salmon fishing, drying, cooking, and eating are an integral part of traditional and contemporary ways of life. It is important to note that the Supreme Court of Canada's recognition of Aboriginal fishing rights includes ceremonial as well as personal consumptive (food) uses of fish. To that end, salmon are celebrated as a foundational part of many Indigenous oral histories, as well as in all forms of art. For example, salmon feature prominently in many of the works by famous Haida artist Bill Reid (Figure 3.2).

More generally, fish and fishing are central to Indigenous cultures across the country. The Assembly of First Nations (2018) describes some of the ways in which fish and fishing are part of celebrating Indigenous cultures and languages:

Figure 3.2. Bill Reid
Source: *Vancouver Sun* (Griffin, 2017). Material republished with the express permission of Vancouver Sun, a division of Postmedia Network Inc.

- Fishing promotes healthy family connections and activities. Fishing is more than the act of removing fish for food – it is teaching and talking about fish, the water sources and the many activities that impact First Nations rights and cultures.
- Fishing in many First Nations is a key activity in transmitting cultures and languages.
- Fishing and food is integral in First Nations cultures. Fishing is an important part of trade, labour and the economy. It helps to shape identity, promote mental, physical and spiritual health, including suicide prevention and life promotion.
- Sustainable, strong fishery economies and water and environmental protection fosters strong individuals and nations. ("Celebrating Fishing, Indigenous Cultures and Languages" section)

Finally, unlike some of the cases that are considered in subsequent chapters, fish are a resource that is ingested, becoming, in a literal sense, part of people's **bodies**. When organized as caloric resources, fish provide a kind of fuel energy for ecosystems such as the temperate rainforests of the Pacific Northwest and for human systems such as the British Empire. However, ingestion also produces vulnerabilities. As we show in more detail in Chapter 6, Indigenous communities that are heavily reliant on fish consumption have been afflicted with mercury poisoning as toxic wastes in the local environment worked their way up the food chain.

Although local availability is one factor that influences food choices, cultural tastes are also susceptible to being governed in other ways. One example is Canada's Food Guide (Health Canada, 2023), a widely used document produced by Health Canada, which in recent years has encouraged the consumption of fish as a lean form of protein. Marketing campaigns by agri-food corporations provide other examples of attempts to influence the evolution of food tastes.

3.5 SUMMARY AND CONCLUSIONS

This chapter has focused on one of the most significant resources in the founding and development of European settlement in Canada. The drive to organize abundant fish in the northwest Atlantic, beginning more than five centuries ago, set in motion a process whereby both fish and human communities became more tightly organized. Fish, conceived as a source of caloric energy or income, provided a resource that was used to establish and grow communities. Those communities, a nexus of human and capital resources, were in turn seen as a resource to be exploited in the pursuit of more fish (and hence more energy, power, and profits). The general trajectory has been one in which the industry has become more capital intensive, with workers as well as fish reduced to abstract units and increasingly governed from a distance. Fish and fishing communities themselves then become resources deployed in power plays at national and global scales.

At the same time, both fish and, in particular, human communities have resisted being organized in these ways and, in so doing, resisted being conceptualized as mere resources. The organization of fisheries was thus not merely imposed from above but can instead be seen as the product of a long series of struggles. These included material struggles over territory, labour, and technology. They also included struggles over ideas, knowledge, authority, jurisdiction, and expertise.

One of the most salient of these struggles was the division between Indigenous and non-Indigenous (or settler) peoples. The restriction of Indigenous fishing rights to a food fishery (or even a food and cultural fishery) relegates Indigenous Peoples to the outside of the mainstream economy, where resources are organized for the pursuit of profit and economic growth. This positions Indigenous Peoples as closer to nature, effectively reproducing a nature–society binary within human society. In this chapter, we have shown how this binary was used to try to exclude Indigenous Peoples from resources that were being organized and accumulated.

The organization of fisheries took place against the backdrop of a social order that was itself the product of struggles to organize other resources. The fishing communities that made the organization of fisheries possible – particularly as these communities became more temporally stable, shifting from seasonal settlements to year-round villages and towns – were unimaginable without the work of domestic reproduction: cooking and feeding, clothing, cleaning, healing, teaching, and so on. Particularly in its earlier phases, this work was mostly unpaid and mostly done by women.

We have shown in this chapter that the organization of fish resources has been an integral part of the story of the building of contemporary Canada. For all of its importance, however, its impact on the built environment has largely been limited to the local scale: the building of fishing communities. For an example of a resource whose organization entailed a dramatic transformation of the environment at the scale of whole landscapes, we turn in the next chapter to the case of forests.

DISCUSSION QUESTIONS

1 This chapter shows how, over time, both fish and fishers have been increasingly organized as resources. What are some other examples of activities where this has occurred? What have been the consequences of that organization? How (and how effectively) was that organization resisted?

2. Fishing rights are clearly an important site of struggle over Indigenous dispossession. How do the channels help us to understand why fishing rights are so significant, both to the project of settler colonialism and to Indigenous Peoples seeking to resist dispossession?
3. What role have governments played in the resourcification of fish? Why might they be motivated to do so?

PEDAGOGICAL RESOURCES

Further Viewing or Listening

A fish tale: The history of the cod [Radio program]. (1997). CBC. https://www.cbc.ca/player/play/1769604492

Historica Canada. (1991). *Heritage Minutes: John Cabot* [Video]. https://www.historicacanada.ca/content/heritage-minutes/john-cabot or https://youtu.be/ds8G9sFOK5w

Markham, N. (Director). (1994). *Taking stock* [Film]. National Film Board of Canada. https://www.nfb.ca/film/taking_stock/

Pauly, D. (2012, February). *The ocean's shifting baseline* [Video]. TED Conferences. https://www.ted.com/talks/daniel_pauly_the_ocean_s_shifting_baseline

Further Reading

Bavington, D. (2011). *Managed annihilation: An unnatural history of the Newfoundland cod collapse*. UBC Press.

Harris, M. (1998). *Lament for an ocean: The collapse of the Atlantic cod fishery*. McClelland & Stewart.

Webster, D.G. (2015). *Beyond the tragedy in global fisheries*. MIT Press.

CHAPTER FOUR

From Forests to Timber

Clayoquot Sound, British Columbia, summer 1993: the site of Canada's largest single act of civil disobedience. Over the course of the summer, more than 800 protestors were arrested for preventing trucks from getting to a clear-cutting site in an old-growth forest in Clayoquot Sound on the west coast of Vancouver Island. Some chained themselves to trucks or logging equipment; some set up barricades to block the road; some camped out in trees destined to be cut. The protests attracted significant international media attention. They also attracted counter-protests by members of the logging community: at one point, 200 litres of human excrement were dumped on the protestors' sites.

The protests were a milestone moment in Canadian environmental history. Never had environmentalism had so much media coverage and generated such intense conflict. The protests demonstrated the power of numbers, of media, and of reaching across the country and world to generate political will. Although logging continues in the area, it has slowed, and in 2000 the area was declared a UNESCO biosphere reserve.

Resourcification saw trees go from forest and ecosystem components to timber, but other kinds of resourcification are at play, too. Trees are tourist attractions and carbon sinks, and they provide a backdrop to what was arguably the most significant environmental protest in Canadian history. Indeed, the characteristics of these old-growth trees – both their size and the roles they play in their ecosystem – made them a desirable commodity for both commercial forestry and environmentalists, and each organized accordingly.

Long before the protest, though, forestry helped build the very communities it divided. The area has been home to the Hesquiaht, Ahousaht, and Tla-o-qui-aht First Nations Peoples for thousands of years, partly because of the abundance that existed in the temperate and biodiverse old-growth forests. More recently, the forestry industry drew more people to the area, increasing the population with the economic draw of steady and well-paid jobs. Accordingly, logging roads were built into the old-growth area, logging permits were sold (and offices to issue them were established), and communities grew up around the logging industry. At the same time, an environmental movement was ignited by the move to protect the old-growth forests, including the trees within them. That movement was a complex alliance between formal environmental organizations such as Greenpeace, some members of Indigenous Nations such as the Tla-o-qui-aht, international celebrities, and local residents, all of whom shared one thing – a desire to stop the clear-cutting in Clayoquot Sound – but who may otherwise have had little in common.

4.1 INTRODUCTION

Forests are an important part of the organizing nature story: people organize forestry, and it organizes them. The development of the commercial logging industry is an excellent example of a case in which ecosystem components – trees – were turned into a resource – timber. Of course, a forest is more than the sum of its timber because a forest also includes soil, berries, shade, fungi, habitat for birds, insects, other animals, and so on. The collective conceptualization of wood as a resource, then, erases many of the non-timber, and thus non-economic, components of forests.

In some sense, the history of forestry in Canada is the history of settler colonialism itself. Wood and timber products were, quite literally, used to construct the railroads, buildings, boats, and homes of settler colonial Canada. Not only was timber used to physically construct the trappings of colonial life, it was also used economically and culturally to support the growth of a particular version of Canada. Economically, the cutting, shipping, and sale of timber (and later, timber products) was foundational to the emergent Canadian economy. Moreover, the transformation of treed landscapes into managed forests symbolized an important element of state-building: the development of measurable, manageable natural resources. In his foundational book, *Seeing Like a State: How Certain Schemes to Improve the Human Condition Have Failed*, James Scott (1998) makes the case that states focus on what he calls legibility in their efforts to establish themselves, build their economies, and manage their populations. By *legibility*, Scott means that states commit themselves to rationalizing and standardizing "social hieroglyphs" (i.e., messy, complex social and natural systems) into administratively more convenient formats (p. 2). In other words, *legibility* means turning the messiness of the world into something that is legible, or understandable, to the state. The belief that states can effectively and should do this is an ideology that Scott calls "high modernism." Scott makes the case that state efforts to make their human and non-human subjects legible are at the core of a great many endeavours, from the development of last names, to urban design, and, most pertinent to our arguments here, to agricultural sites and forests.

Scott (1998) uses the example of German forestry to make this point. Specifically, he documents the development of scientific forestry in the late 1700s in Saxony and Prussia (both in what is now Germany). These scientific foresters first aimed to quantify and measure the economically viable wood within a plot of land. Following from this quantification, they aimed to optimize, cutting down treed areas to replace them with monocrops of easily harvested, fast-growing, economically valuable timber. In sum, they measured, quantified, and organized nature for economic optimization. One consequence of making forests legible in this way, and controlling or getting rid of forests' "messiness," is a dramatic reduction in forest biodiversity. When managed in this way, forms of forest life are only sustained if they contribute (or at least do not diminish) economic value.

Thus, going from forests to timber is far more than a semantic change: it is a change in the way in which particular tracts of land are conceptualized and manipulated or managed. As discussed earlier, forests consist of trees, shrubs, ferns, soil, mushrooms, and a variety of animals, whereas timber consists of tree trunks of either hardwood or softwood. The difference between forests and timber is also addressed by William Cronon (1983) in his book *Changes in the Land*, which documents changes in the social and ecological landscapes of New England during the period of European settlement. Cronon describes how Europeans arriving in New England viewed the landscapes, and particularly the forests, as sites of abundance, because most forests in Europe had by that time already been severely depleted. Moreover, because the Indigenous residents of the land did not own their territory in terms of European systems of property ownership, this abundance appeared free for the taking, and forests became timber, as wood was rolled into systems of capital and commodified for export and construction.

Although Cronon (1983) and Scott (1998) were writing about the United States and Europe, similar stories played out in Canada. In this chapter, we explore the relationship between people and forests and timber, exploring the history of forestry in Canada, the role of forest protection in driving a new wave of Canadian environmentalism, and a new form of forest monetization: the development of forest protection through climate change mitigation mechanisms such as carbon markets. As with the rest of the chapters, we explore these three case studies first and then address their application in the various channels through which resources are organized.

4.2 GROWTH OF TIMBER: SAINT JOHN, NEW BRUNSWICK

The opening story in this chapter was focused on the West Coast of the country, but forestry has been equally important to settlement and organization on the East Coast. Basic geography meant that timber (and other resources) extracted from inland Canada could be shipped out of East Coast ports directly to Europe. The development of ports brought with it the shipbuilding industry, and the combination of logging and shipbuilding played a significant role in the colonial settlement of the East Coast.

This was especially true in New Brunswick (Map 4.1). Cole Harris (2009) traces how early European settlement in Atlantic Canada was especially concentrated in areas where inland rivers met or emptied

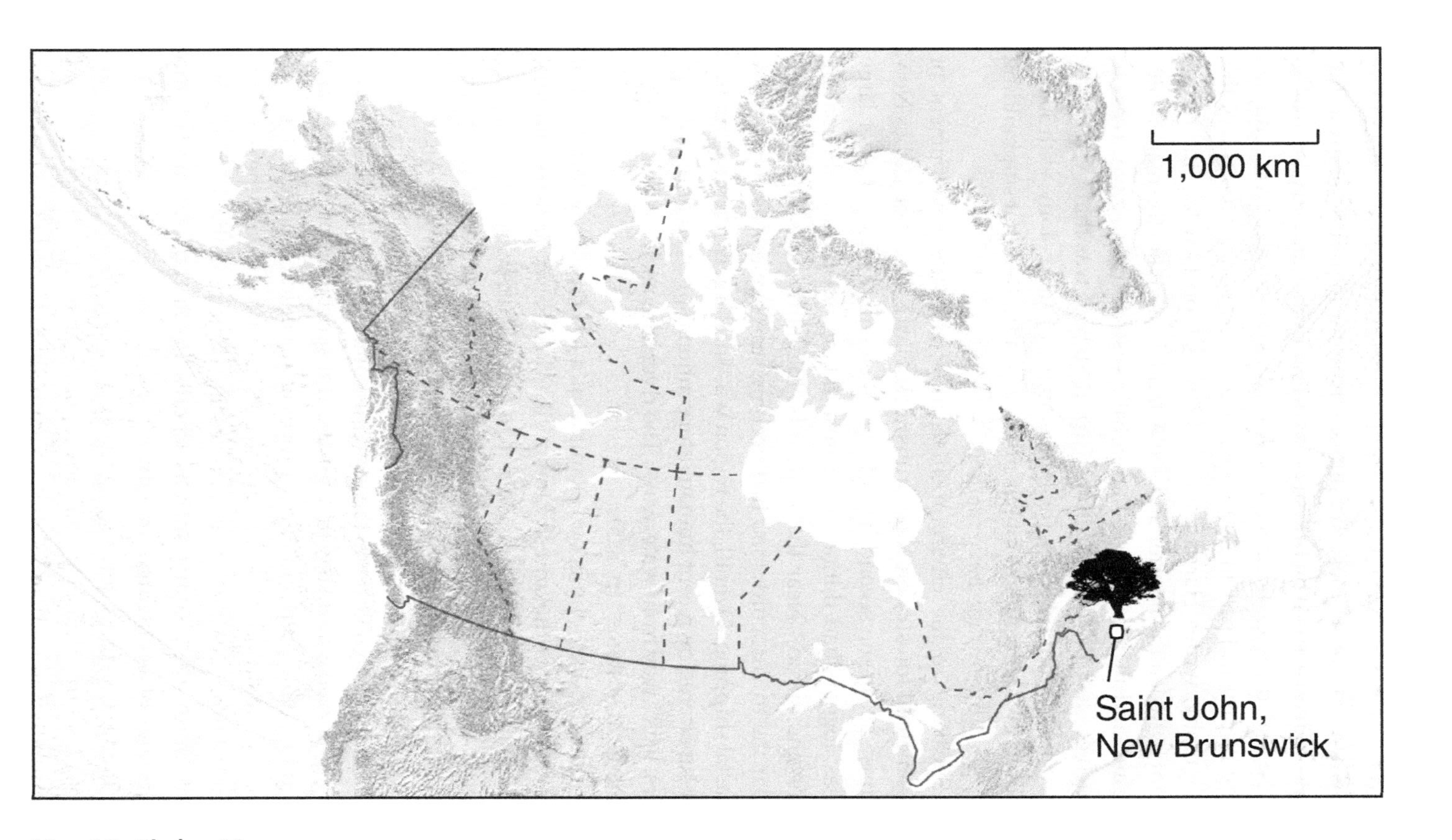

Map 4.1. Timber Map

into the ocean because it was (relatively) easy to fell trees, float them downriver to port cities, and bundle them for export on ships to Europe. Indeed, driven by British demand for timber, forestry intensified in New Brunswick because of the region's forest resources, coupled with its physical proximity to ocean transportation. As a result of this geography, writes Harris, New Brunswick's ports were "the hinge-point between the interior and the Atlantic; the rivers the main routes inland, and their tributaries a dendritic system that ... gave access to much of New Brunswick's forest" (p. 188). So it was that by 1851, the port city of Saint John was the largest in the British colony, with 30,000 people: "whatever Saint John is, it must be admitted that shipbuilding and the timber trade have made it" (p. 188).

Speaking to the environmental impacts of the development of commercial forestry on New Brunswick, Harris (2009) describes New Brunswick in the second half of the nineteenth century: "Over large areas, loggers had thinned the forest, taking out most of the mature pine and spruce. Near the shipyards, the hardwoods had been cut. Near the towns, cordwood had become scarce and was supplied by coastal shipping. Overall, the region was still largely forested, but forests unaltered by recent human interventions survived only in its corners, well away from waterways. The rivers themselves were changed ecological systems" (p. 225).

Indeed, Saint John is one example of a town built by forestry, but there are many other forestry towns like Gold River, British Columbia; Fort Frances, Ontario; and Miramichi, New Brunswick. An analogous story can even be told about the community situated at the intersection of the Ottawa and Rideau Rivers. Known first by its Algonquin name, Adawe (meaning "trade"), and then by the colonial names of Bytown and then Ottawa, the growing community depended heavily on the forestry industry, first for the transportation of raw logs and then to enable pulp and paper processing (Benidickson, 2010). In Bytown (Ottawa), as in Saint John and other forestry towns, the development of timber extraction and processing came with the development of community amenities such as schools, hospitals, retail, and all the trappings of a contemporary city.

Today, Canada's forestry industry is a significant economic engine. In 2015, the sector employed more than 200,000 people and contributed more than $32 billion annually to Canada's GDP (Export

Development Canada, 2017). As significant as those numbers are, they pale in comparison to what they used to be: in the 1800s, nearly half of Canada's male population worked in the forestry industry (Natural Resources Canada, 2017). Given its historical importance, it is perhaps unsurprising that the forestry industry has been one of the primary organizers of Canadian society: the industry established towns, funded governments, and provided jobs for years. As the next case studies show, though, turning forests into timber is only one way in which forests can be organized as resources.

4.3 TREES, NOT TIMBER: PORT RENFREW, BRITISH COLUMBIA, AND DARKWOODS

In this section, we discuss two cases in which non-logged forests are conceptualized as resources. This contrasts with the previous section, in which we discussed the physical extraction of certain components (trees and the timber derived from them) from the forest ecosystem. Here, intact forests provide different kinds of resources: a spectacle or experience that can be consumed and ecosystem services such as carbon sequestration (to mitigate climate change). In both cases, forests are turned into commodities without trees or other components being physically removed. In reviewing the two case studies that follow, we ask you to consider whether this resourcification of forests is a move toward more just and sustainable futures or whether it is commodification in a new outfit. Can commodification and sustainability coexist?

The community of Port Renfrew, home to just fewer than 200 people, sits on the west coast of Vancouver Island (Map 4.2). A two-hour drive from Victoria, the community has branded itself as "Canada's tall tree capital," where tourists come to visit and walk through some of Canada's oldest-growth trees. The region boasts some of the oldest and tallest trees on the continent: the (in)famous 1,000-year-old tree named Big Lonely Doug (pictured in Figure 4.1) was a seedling "around the time the Norse explorer Leif Ericson was building sod houses in what is now Newfoundland" (Rustad, 2016, para. 22).

This focus on ecotourism is a dramatic departure from the town's history, which, as the town's website notes, was "long a forestry and

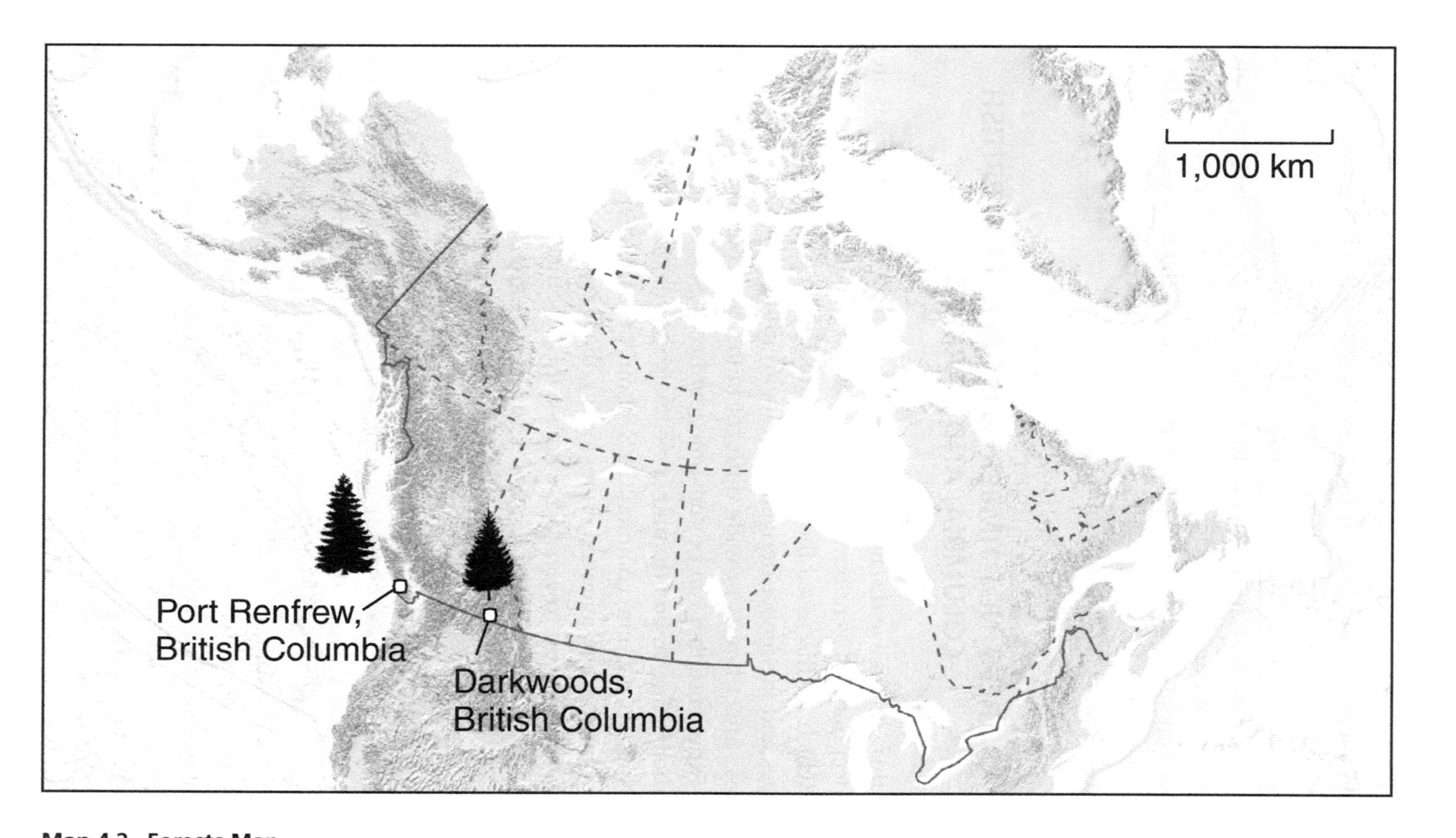

Map 4.2. Forests Map

Figure 4.1. Big Lonely Doug
Source: Photograph by TJ Watt – Ancient Forest Alliance. From the *Walrus Magazine*, https://thewalrus.ca/big-lonely-doug/.

commercial fishing hub" (Port Renfrew Chamber of Commerce, n.d., para. 1). Indeed, as with Saint John and other towns discussed in the preceding section, logging was the primary interest of the Europeans who first settled the region (in the 1920s), who then built rail connections to facilitate the export of local timber. So it was a complete 180-degree turn when, in May 2019, the town's Chamber of Commerce publicly decried the BC provincial government's plan to log the area. Why? Because the trees were worth more alive than dead. As Chamber of Commerce President Dan Hager told a reporter, "[Preserving the trees is] a lot better than cutting them down, because you cut them down once, you run them through the sawmill, they build somebody's deck and that's it. But, if you leave them standing, people come over and over again to look" (Kines, 2019, para. 17). In the same article, Hager also notes that "those tourists have money. They bring money and the more of it that we have in the immediate driving area of Renfrew, the better it is for our local economy" (Kines, 2019, para. 16).

For Hager, the forest is still a resource, but in this case the resource is a source of tourism revenue, not timber. Moreover, this form of monetization has been facilitated by Hollywood and pop culture: two of the protected areas are called Mossome Grove and Avatar Grove, the latter named after director James Cameron's environmentally themed 2009 Hollywood blockbuster film.

The Port Renfrew case is an example of how, in recent years, economic benefits have been associated with not cutting down forests. In Port Renfrew, the trees are still conceptualized as a resource, although one that is not to be extracted but that needs to be sustained to continue generating value. The next case we discuss provides an example of forests as sources of ecosystem services (Box 4.1).

BIG IDEAS IN SMALL BOXES

BOX 4.1. ECOSYSTEM SERVICES

Ecosystem services are an important concept in environmental governance. The term has multiple definitions, but the central idea is that ecosystem services are those things that nature (broadly defined) provides for free. In their foundational 1997 article, Robert Costanza et al. define ecosystem services "as the benefits human populations derive, directly or indirectly, from ecosystem functions" (p. 4). The article's authors list seventeen broad categories of ecosystem goods and functions. The categories are diverse and range from climate regulation and waste treatment to pollination, food production, and sites for recreation. Of course, each of these categories can be broken down even further, and the barriers between them are fuzzy. For example, food production is not a stand-alone category because it depends on other categories of services such as soil formation, climate regulation, water supply, and pollination. Nevertheless, Costanza et al. calculate the approximate value of the world's ecosystem services at US$33 trillion annually. Throughout the article, they repeatedly note the limitations of their study, including stating that "the real value is almost certainly much larger" than US$33 trillion and that putting a dollar figure on the earth's services is impossible, given that "many ecosystem services are literally irreplaceable" (p. 259).

Despite these limitations, the notion of calculating the economic value of ecosystem services has taken off, with valuation being used to, for example, put a price on carbon, fishing quotas, or other activities that cause damage to ecosystem health and function. Indeed, since Costanza et al.'s (1997) article, research and action with respect to ecosystem services has ballooned. At the time of writing, a Google Scholar search for the term *ecosystem services* turned up more than one million hits.

The uptake was not only academic. Ecosystem services were a cornerstone of the United Nations' important Millennium Ecosystem Assessment (2005) and are used by many non-governmental organizations (NGOs) to make the case to protect ecologically sensitive areas. As Kull et al. (2015) note, much of the excitement for the concept arose from conservationists' ongoing frustrations with government inaction on policy grounds alone; the possibility of engaging market actors in ecosystem protection presented the possibility of renewed energy, focus, and financing directed toward environmental protection. In the terms that we use in this book, it appeared to be a way to use the economies channel to make the case against resource extraction and overuse. Yet despite the promise and excitement of ecosystem services as a panacea for environmental woes, challenges have arisen. Putting a price on ecosystem services can reinforce unequal power relationships and lead to social injustice, because some actors have much more financial resources than others (Kull et al., 2015). It can lead to moral licensing – the phenomenon wherein actors who pay for ecosystem services then feel they have licence to pollute or use even more. Also, as Constanza et al. (1997) suggest, it may produce futile attempts to price things that have no price because they are irreplaceable: what is the price of breathable air or drinkable water?

Forests provide a variety of ecosystem services, from shade to habitat, but the service that is arguably at the forefront of current political discussions is their service as carbon sinks. Because trees absorb carbon dioxide (CO_2), they are considered an important tool in addressing climate change. In a video released in September 2019 starring climate activist Greta Thunberg, George Monbiot explains the importance of trees in addressing the climate crisis: "There is a magic machine that sucks carbon out of the air, costs very little and builds itself. It's called ... a tree" (Conservation International, 2019). Because of trees' ability to, as Monbiot says, lock carbon away, investing in standing, healthy forests is seen as an important way to address the climate crisis. This shift is an important one: whereas once the primary site where forests interacted with economies was in the forestry (logging) industry, now it increasingly happens through the recognition of forests' climate absorption capacity in the context of a rising (and justified) panic about climate (in)stability.

A good example here is the work of the Nature Conservancy of Canada (NCC). The NCC is the largest land conservation organization in the country (NCC, n.d.-c). Since its inception in 1962, the NCC has protected more than 35 million acres across the country. It uses donations to purchase or lease ecologically sensitive land and to treat it in ways that are consistent with the organization's conservation goals. One of the NCC's signature projects is Darkwoods, a 155,000-acre conservation area in British Columbia's west Kootenay mountains (Map 4.2), described by NCC (n.d.-a) as follows: "Conserved in 2008 and expanded in 2019, Darkwoods spans 63,000 hectares of remote valleys, mountains and lakes, providing essential habitat for dozens of species at risk. The conservation area plays a central role in a network of parks, wildlife management areas and conservation lands that encompass over 1,100 square kilometres" (para. 2).

Most relevant to this chapter, Darkwoods generates carbon credits by not being logged. Before its purchase by the NCC, Darkwoods had been logged on a small scale for four generations. Since the property was purchased by NCC, carbon credits are generated by abstaining from logging: because the forest is allowed to grow in the absence of logging, tree and plant growth removes CO_2 from the atmosphere. To the extent that this CO_2 removal mitigates climate change, it provides economic value. Needless to say, getting an accurate assessment of the value created by not doing something is a difficult and contested process. Third-party organizations such as the Rainforest Alliance and Verified Carbon Standard – separate from organizations such as Darkwoods that seek to generate and sell carbon credits – have emerged to audit and approve the measurement of carbon credits. Together with other carbon offset projects, this activity represents a growing sector of economic forest activity. As NCC puts it, it is a way of using the "power of the carbon market and conservation finance to advance large-scale conservation in Canada" (NCC, n.d.-b, para. 1). In other words, it taps into Canadians' sense of forest stewardship and couples it with economic activity and the climate crisis.

We address the incorporation of biophysical function into market systems in Chapter 8, but the key point here is that forests have, in turn, been turned into sites of profit through forestry, ecotourism, and carbon sequestration.

4.4 CHANNELS IN ACTION: ORGANIZING FORESTS

If institutions are permanent solutions to permanent problems, then the bevy of forestry institutions hints at a suite of problems: infestations, such as the mountain pine beetle; loss of old-growth forest; and international trade disputes are but three examples. Although the problems vary region by region and change over time, the theme remains the same: forestry institutions are primarily concerned with the problem of how to convert wooded landscape into resources and with solving the challenges that arise in the conversion process. That is, when it comes to forests and forestry, the permanent problems with which government institutions are concerned are the problems of commodification.

We can return to Scott's (1988) discussion of forest commodification in Germany, mentioned earlier. An important part of this story is the transformation of the forest itself, from a messy tangle of mixed ages and species of trees to orderly rows of single ages and species of trees that are easier to measure and harvest. Although it is perhaps most obvious in the case of timber extraction, even the understanding of forests as tourism sites or carbon sinks requires the imposition of a certain degree of orderliness on the landscape. In addition to measuring and harvesting, orderly forests provide a way to more accurately predict future revenue and to harvest within the Maximum Sustainable Yield (MSY): that is, the maximum number of trees that can be harvested at one time without diminishing future years' harvest, similar to the idea of total allowable catch discussed in Chapter 3. MSY also presumes that trees are merely containers for interchangeable units of lumber. The idea that individual trees, such as Big Lonely Doug, mentioned earlier, have value beyond a quantity of timber (or even carbon storage capacity) cannot be seen through the lens of MSY.

A variety of institutions have arisen to manage the complex mix of MSYs (and their ever-evolving science), forestry licences, public opinion, Indigenous treaty rights, international commodity prices, and so on. As shown in the previous chapter, one can see how several different types of institutions organize – and are organized by – forests, and, also as in Chapter 3, we can begin with **government** departments. The Canadian Constitution specifies that provincial governments "may exclusively make laws in relation to ... *development, conservation and management of* non-renewable natural resources and *forestry resources in the province, including laws in relation to the rate of primary production therefrom*" (*Constitution Act*, 1982, s. 92A 1 [b], emphasis added). In other words, the provinces are responsible for managing forestry activities within their boundaries. That said, the federal government also plays a role. Federal lands, including national parks and Indigenous reserves, fall under the jurisdiction of the federal government, and federal laws such as the *Species at Risk Act* (2002) and the *Fisheries Act* (1985) both affect forestry because they limit or prohibit forestry activity in sensitive areas. In addition, only the federal government can make international agreements, so the federal government affects forestry activities by signing on to international environmental treaties such as the *Convention on the International Trade in Endangered Species of Wild Fauna and Flora* (CITES; 1973). The federal government also oversees navigation of the ongoing softwood lumber trade dispute with the United States.

Meanwhile, provincial governments, for their part, make decisions about allocating forestry licences, allowable cut blocks, and so on. As with the fisheries (see Chapter 3), privately owned entities compete, but in the context of government regulation that determines how much can be cut, where, and by whom. All provinces have a ministry responsible for forests, although few have a stand-alone forestry department. For example, Ontario's Ministry of Natural Resources is responsible for forestry but also for hunting, fishing, provincial Crown land, and wildlife. Similarly, Manitoba's Ministry of Sustainable Development is responsible for forestry activity in the province; it is also responsible for fish and wildlife, water, waste management, and biodiversity. New Brunswick has a Department of Energy and Resource Development, which, under its Natural Resources section, has responsibility for forests and Crown lands. Although the ministry names differ across provinces, a commonality is that forests are seen as resources to be developed. The point here is that Canadian provincial governments have organized themselves around the forestry industry, that is, that forests have been organized by governments, and governments have been organized by forests.

Another way in which forests and Canadian institutions have mutually shaped one another is through the courts. Forests have long been a focus of legal battles. One of the most notable examples is *Haida Nation v British Columbia (Minister of Forests)* (2004), which hinged on forestry but was more broadly about governments' obligations to consult with Indigenous Peoples. In this case, the government of British Columbia transferred a forestry licence on the Haida Gwaii Islands to Weyerhaeuser Company Limited without first consulting with the Haida Nation. At the time of the transfer, the Haida Nation had a pending but unresolved claim to land title on the contested land. The Supreme Court of Canada ruled that the Crown has a duty to consult in good faith with Indigenous Peoples on use of their land – a ruling that has shaped how government institutions operate. The case was specifically about forestry, but it established a precedent for the legal duty to consult that operates well beyond the forestry sector. In this case, disputes about how forests can be used or interacted with have been instrumental in shaping Canada's legal landscape, which in turn reshapes forests, as an increasing number of stakeholders are involved in land use decision making.

Undoubtedly, these government institutions organize forestry resources: on the basis of court decisions, government agencies issue logging permits (or not) to forestry companies, they work with private industry to plan for the planting of new trees, and they oversee the application of relevant legislation. Together with private industry, government institutions are the mechanism through which forests become organized as timber resources, an entity that looks and functions quite differently.

Looking beyond governments to other public-sector institutions such as universities, one can begin to see how forests are organized through the channel of **culture and ideas**. We return to our discussion of culture later in this section, but here we consider how institutions that generate knowledge about forests have been organized and how different forms of knowledge are mobilized to organize forest resources in particular ways. In Canada, most universities are public-sector organizations that operate at arm's length from government. Provinces with a prominent forestry industry have developed forestry management programs at some of their universities: the University of British Columbia in Vancouver; Lakehead University in Thunder Bay, Ontario; and the University of New Brunswick in Fredericton offer the country's leading forestry programs, and these three provinces are the ones with the most productive timber industries. These universities were (like most in Canada) established and in part funded by provincial governments, although they have considerable autonomy from government in their day-to-day operations and in the longer-term management of their programs.

Industry associations provide another institution, at even greater remove from government. Organizations such as the Canadian Forestry Association (CFA) and the Forest Products Association of Canada (FPAC), are voluntary associations made up of corporate members of the forestry industry. Their primary purpose is to act on behalf of the forestry industry, so they work to promote forestry activity (including by lobbying government) as well as to educate about the role of forestry in the country. Perhaps unsurprisingly, organizations such as FPAC work to dispel what they see as stereotypes of their industry and posit themselves as social and environmental leaders. For example, FPAC has a job-matching program (#TakeYourPlace) that aims to "tap into additional talent pools of Canadian women, Indigenous Peoples, and new Canadians to grow our workforce" (FPAC, 2019, para. 1). Promoting itself as an environmental leader rather than a laggard, FPAC has also committed to removing 30 megatonnes of CO_2 from the atmosphere by 2030.

In a different vein, the Forest Stewardship Council® (FSC®) has a vision to "ensure that forests are managed in ways that are socially

Figure 4.2. Forest Stewardship Council Logo
Source: Courtesy of the Forest Stewardship Council®.

beneficial, environmentally appropriate, and economically viable" (FSC, 2020, p. 5). FSC is a third-party certification organization that sets standards for forestry management, and it certifies particular products as meeting those standards. Consumers can look for the FSC logo on a variety of goods – everything from lumber to toilet paper to printer paper, even this book! – to choose to purchase a product meeting those standards (Figure 4.2). Here, we have an example of a private organization (FSC) mediating the relationship between producers and consumers, facilitating customer choice of products deemed more sustainable. Of course, FSC as an organization struggles with the same challenges as other third-party certification organizations: complete objectivity or neutrality in determining what is environmentally appropriate or socially beneficial is impossible, but third-party certification can provide an assessment that is relatively independent from the producers who have an obvious stake in such assessments. It can be a fine line between having standards so high that no company or product can meet them and standards so low that the certification is not meaningful. Nevertheless, third-party forestry certification is another example of mutual co-organization: how and what consumers buy affects forest makeup, and organizations such as FSC hope to change consumer behaviour and, in turn, how and where forests grow.

All of the institutions just mentioned – universities, industry associations, and third-party certification organizations – as well as government departments, Indigenous groups, and environmental NGOs such as Greenpeace, Ecojustice, or Friends of Clayoquot Sound – engage in the generation and dissemination of ideas about forests. In some cases, these knowledges about forests facilitate their being understood as resources, whereas other forms of knowledge work against this understanding. Still others – ideas about forests as sites that generate commodifiable ecosystem services – may be a more complex mix of the two.

If one considers **economies**, one can see that markets for forestry products are clearly another important channel for the organizing of forest resources. If economies are about who gets what, then forests are a central nexus through which the nature–money relationship is mediated. The transition from ecosystem component to resource is, of course, about money: who gets it, who spends it, and how.

Provincial governments are in a tough spot when it comes to forestry. On one hand, provincial governments can benefit from issuing (and receiving payment for) forestry permits. Of course, the benefits are not universally shared, as the Clayoquot Sound example that introduces this chapter shows. Given the prevalence of forestry communities across the country as well as the fees that accrue to provincial governments when forestry permits are issued, one can see that provincial governments have real incentives to encourage, rather than limit, forestry activity.

At the same time, provincial governments also have the responsibility to protect forests for their various ecosystem and cultural values, which do not necessarily deliver an immediate economic return. They also have a legal obligation to act in a way that is consistent with the constitutional imperative to recognize and affirm the rights of Indigenous Peoples, and many of those rights are directly or indirectly tied to forests. Moreover, standing forests provide more diffuse economic value through the ecosystem services discussed earlier, as well as through tourism and increased real estate value. Homes backing onto forested areas are more valuable, as are homes with views of and access to green spaces.

So here is another example of mutual co-organization, only this time through economies. In one direction, economies organize forests in a very concrete and material way: commercially planted lots look different than old-growth forests – they consist of straight rows of trees of the same age and species, rather than a mix of ages and species. Aside from the forests themselves, forestry economies have organized people geographically into forestry towns, occupationally into forestry workers, and educationally through training centres for forestry professionals. The focus on exporting timber turned many coastal **communities** into booming port towns, and forestry revenue was – and to some extent continues to be – a force in driving the establishment and continued existence of many communities. In Chapter

3, we discussed a shift from geographically specific fishing communities (plural) to a more corporate, less place-defined fishing community (singular). A similar evolution can be seen when it comes to forests: a shift from multiple towns identified as distinctive forest(ry) communities to a singular forestry community that is occupationally rather than geographically defined and increasingly represented by industry associations such as the CFA and FPAC.

Historically, as a significant employer, a force driving the creation of many communities, and an economic bridge both to Europe and across the country by way of the then-newly developed railroad (see Box 7.2), the forestry industry was an important thread in the emerging fabric of a colonial Canadian identity. Some of the **cultural** forms that were part of this colonial identity persist to this day.

For example, the development of the forestry industry gave rise to the lumberjack (a Canadian term that was first introduced to the *Oxford English Dictionary* in 1831) and the lumberjack coat or lumberjacket – the black-and-red checkered jacket – that continues to define stereotypically (male, working-class) Canadian fashion – often satirically, sometimes seriously. Historically, large numbers of men employed as lumberjacks and log drivers (men who rode rafts of timber downriver out of the forests and to shipping ports) were part of an emergent colonial identity that tied masculinity to resource development. Geographer Michael Ekers (2009) writes that in the Great Depression of the 1930s, the chief forester proposed forestry development projects as a solution to "tying up our two greatest problems – forestry and young men's unemployment" (Ekers, 2009, p. 309). The lumberjack figure features prominently not only in the colonial frontiersman image of Canadian settlement, but also in the contemporary commercial value seen in harkening back to "old timey" Canadian imagery. Molson's series of "I am Canadian" beer advertisements, featuring Joe Canadian wearing the iconic lumberjack coat, is one example. Acadia University's mascot (the Axeman) is another. Whether used to sell beer or cheer on athletic teams, the image of the lumberjack has a cultural resonance that extends well beyond the resource economy from which it originated.

So intertwined were forestry, masculinity, and colonial culture that in 1979 the National Film Board produced a short vignette called "The Log Driver's Waltz" as part of its cultural vignettes series – a collection of short films (five minutes or less) to be screened between shows or commercial breaks (Weldon, 1979). The series received a $2 million budget, as part of a $13 million government commitment to federal cultural agencies to promote national unity. The vignettes were shown during prime time on CBC and in children's programming slots, and they were then picked up by CTV, Global, TVOntario, and other broadcasters. This series – and the "Log Driver's Waltz" in particular – are a fascinating example of the Canadian state dedicating financial resources to the promotion of a particular idea of resources and cultural identity.

The song that accompanies the film is sung by Kate and Anna McGarrigle and is the sonic backdrop to the three-minute vignette. The vignette begins with black-and-white footage of log drivers at work, which transitions to a cartoon of log drivers performing jaunty dance moves as they float downriver on logs for a passing audience of moose, beavers, and adoring young women (if you have not seen it, it is worth watching; Weldon, 1979). Every person in the clip is white, and the earworm chorus is as follows:

> For he goes birling down and down white water
> That's where the log driver learns to step lightly
> Yes, birling down and down white water
> The log driver's waltz pleases girls completely

The plot of the song revolves around a young woman who is quite taken with the log drivers and their dancing (i.e., the balancing act of staying atop rolling logs on an active river), much to the disapproval of her parents ("To please both my parents, I've had to give way / And dance with the doctors and merchants and lawyers"). In the final verse, she is vindicated:

> Now I've had my chances with all sorts of men
> But none as so fine as my lad on the river
> So when the drive's over, if he asks me again
> I think I will marry my log driver

Much could be said about this clip with respect to gender, race, Indigenous dispossession, and so on. In the interests of space, we focus on how forestry has been both organized and an organizer, most obviously of economies and communities but also of institutions and identities, and on the role of pop culture (i.e., this video) in promoting the nature–society binary. With respect to **identities**, the video speaks to William Cronon's (1995) points about the valorization of the frontier experience that wilderness makes possible, because the physicality of the log driver's work and his more immediate connection to nature makes him a more attractive partner than doctors and merchants and lawyers. In this song, the physical work of organizing forest resources (log driving) exemplifies a particular masculine identity that is highly praised by the song's narrator. For Cronon, conceptions of masculinity that were dominant among (North) American settlers, particularly in the nineteenth century, saw value not in wilderness itself, but in the encounter with wilderness that was only possible on the frontier. Masculinity could be asserted by taming the wilderness, making it suitable for civilized settlement by women, children, and others – urban professionals, racialized minorities – who could be looked down on as less masculine.

To this day, the forest is often framed as a place of uninhabited wilderness – a place devoid of humans. This is consistent with a popular **culture**: the forest is the scary place in fairy tales where only monsters and animals live and, more recently, a place where people go to rejuvenate themselves (hence, *re-creation*) by escaping the stressful trappings of modern life. This shift of forests and wilderness from a place of fear to a place of both recreation and extraction and profit is an important one, and one that is reflected in the project of an emergent national identity.

Indeed, depicting forests as wilderness was a central part of the emergent colonial Canadian identity. John O'Brian and Peter White (2007) describe this phenomenon in their book about the Group of Seven, a collection of Canadian artists active in the 1920s and known for their paintings of the Canadian landscape – most notably the Precambrian Shield in northern Ontario and, to a lesser extent, the mountains of British Columbia. Their images typically depict a combination of trees, sky, and water from different perspectives and in different seasons. What is not depicted, however, is just as important: "Devoid of human presence, [the] works also eschewed the industrial development of the land that often existed just outside the picture frame" (O'Brian & White, 2007, p. 14). The immense popularity of the group's work, despite its selective representation of Canadian wilderness, derives not from its artistic merit but from what such depictions represent. One theory is that the group's popularity rose in direct proportion to "the growth of crowded industrial cities and the simultaneous development of easy means to escape them" (Cole, 1978, p. 69). Another is that the depictions of Canadian wilderness spoke to Canada's emergence as a nation independent from Britain. As O'Brian and White (2007) note, "As a relatively new nation-state that was still very much attached through its imperial connection to Britain, Canada was actively in need of signs and symbols with which to assert a distinctive national identity. If considered in relation to the refined, picturesque English landscape, the Precambrian Shield could readily constitute such a vehicle of national meaning and feeling in Canada" (pp. 14–15)

In other words, the landscapes painted by the Group of Seven, which depicted scenes without any human presence, were seen as representing something distinctively Canadian. Decades later, popular culture looped back again to the contrast between Canadian identity found in the forest and the more "refined" alternatives with the "Log Drivers' Waltz." The forest here, however, unlike in Group of Seven paintings, is no longer uninhabited by humans. Rather, it is a site from which resources are being extracted by a particular kind of man: the log driver. The fact that the National Film Board dedicated its resources to immortalizing and promoting log driver culture speaks to the importance of resource extraction as part of mainstream Canadian identity at the time the film was produced.

Forests and wilderness, then, have been central to the development of a colonial Canadian **identity**. Indeed, as Baldwin (2009) notes, "The concept of wilderness enjoys the dubious distinction of being one of colonialism's most enduring symbols in Canada, an empty space, devoid of humans ... [and] quite literally founded on the erasure of aboriginality" (p. 432). Forests are an important part of the story about the simultaneous rise of wilderness and erasure of Indigeneity because they are the site of the story. Forests are a place in which the story plays out: that part of the landscape that was depicted as void of human habitation so it could be filled with the colonial priorities of marketization (i.e., reconceptualizing the forest as timber and recreation).

To be sure, given the importance of timber for building construction, European settlers organized (and continue to organize) forests as resources with a view to transforming the **built environment**. As we have noted, however, forests were not uninhabited, or untransformed for human use, when European settlers arrived. For thousands of years, many types of Indigenous architecture have been made possible through the existence of particular kinds of trees in particular places. For example, the plank houses built by the groups living on the coast of what is now British Columbia – the Coast Salish in the south; the Kwakwaka'wakw, Nuu-chah-nulth, and Nuxalk on the central coast; and the Haida, Tsimshian, Gitksan, and Nisga'a in the north (Mills & Kalman, 2007) – are made from the enormous cedar trees unique to that ecosystem's temperate rainforest. Because the trees were large in height and circumference, plank houses were large enough to accommodate the members of an extended family in one house. Because cedar is strong, light, and straight, it had a variety of practical uses, including totem poles, masks, and longhouses. In addition to being functional, the houses and their structure were also built to denote the home's occupants: "Carved house frontal poles would occasionally be positioned at the entrance, particularly amongst the Haida and Tlingit. These poles typically depict the crests and lineage of a family, as well as the hereditary rights and ancestors of the owners. Many First Nations decorated house posts, mortuary poles, and memorial poles with intricate carvings of stylized human figures and animals" (Huang, n.d., "Uses of Cedar" section).

Cedar, then, in its abundance and with its particular physical qualities, not only provided (and continues to provide) functional structure for many communities along the Pacific coast, but it was also a part of organizing societies in that region: enabling extended families to live together in larger homes, allowing for permanent settlement, and making possible certain culturally significant practices.

Finally, we have already discussed how forests have shaped and been shaped by identities (in particular, gender and national identity). As we described in Chapter 2, though, the **bodies and identities** channel also has a material dimension: physical bodies themselves. In the case of the Clayoquot Sound protests, human bodies mattered. They mattered first because they were the weapon of choice for protesters, who put their bodies in front of trucks and on top of trees to prevent forestry operations from proceeding. When protesters sought to generate popular support for their actions, however, by circulating images for a largely urban Canadian public, human bodies were conspicuously absent from their framing of the forest. As scholar David Rossiter (2004) notes in his work on how the Clayoquot forests were portrayed during the protest, Greenpeace campaign materials "contrast[ed] stunning, colourful images of seemingly untouched watersheds and valleys against pictures of scattered, broken forest landscapes resulting from clear-cut harvesting[. In doing so,] Greenpeace campaigners are able to provide in their publications instant visualizations of the impacts of industrial forestry. To most observers the intact landscapes represent what is natural, what should be" (pp. 144–5).

This is a critical point: protestors at the site relied on the erasure of humans from the landscape – that is, the wilderness factor – to evoke sympathy for their cause. Like the Group of Seven painters almost a century earlier, they relied on a distorted depiction of the landscape they were representing. In the case of the Clayoquot protests, these choices were made to appeal to an audience that had already been acculturated to understand nature and wilderness in particular ways. As Rossiter (2004) puts it, "Metropolitan conceptions of nature leave no room for work in the woods and instead find in untouched nature an opportunity for leisure based on play and aesthetic appreciation" (p. 148). Erasure of bodies from the wilderness landscape is, of course, not unique to forestry, but it is certainly a feature of forests' depiction in the cultural imagination.

This dual use of bodies is an important part of understanding the ways in which ecosystem components become resources. Bodies are removed from representations of the landscape to draw public sympathy, but this, perhaps ironically, also removes from view the loggers and their economic activity. The resulting perception that Canada's forests are somehow a pristine wilderness presents a contradiction: it gives the impression of responsible forest stewardship while also facilitating ongoing forestry activity. It also enables the kinds of protection activities embodied by the Port Renfrew and Darkwoods examples, where imagery of a pristine forest becomes an economic resource, for tourism in the Port Renfrew case and ecosystem services in the Darkwoods example.

4.5 SUMMARY AND CONCLUSIONS

From the maple leaf on the flag to maple syrup, the "Log Drivers' Waltz," the Maple Leafs hockey team, and the iconic Group of Seven paintings, forests and trees have been knit into the Canadian colonial identity. Yet the ways in which people benefit from them are not always as forests and trees: timber, and the forestry industry more broadly, have been a cornerstone of Canadian politics and identity. As we reviewed in this chapter, many communities and economies have relied on the timber trade for their existence, and wood has been instrumental to the physical construction of the landscape – from Indigenous Peoples' use of cedar to the use of timber in the construction of the Canadian National Railroad. But the representation of forests more broadly has been complex and contested.

In this chapter, we have attempted to separate out the idea of a forest from the idea of timber and, more important, to identify why this difference matters. We suggest it matters partly because the idea of forestry resources has become baked into common discourse without giving thought to the channels through which forests become resources and what effect that has on politics. Moreover, because referring to forests as forest resources renders invisible the transformation from forests to resources, the cultural, political, and economic effects of the transformation are also invisible, making it seem as though forestry communities, culture, and institutions have somehow always been there rather than the reality that they are a product of resource development and an embodiment of the mutual co-organizing of people and place.

DISCUSSION QUESTIONS

1 What ideas or images come to mind when you think about forests? Are forests places that are relaxing or exciting? Dangerous or inviting? Exotic or banal? How have the various channels shaped this way of thinking about forests?

2 The idea of forests as resources has broadened over time, from a stricter emphasis on timber resources to a broader conception of forests that includes value as recreational resources and repositories of ecosystem services. Is this broadening a welcome move toward greater sustainability, or does it further entrench the commodification of forests and the dominance of resource thinking?
3 How might people think about, and act in, forests in ways that do not reproduce the nature–society binary?

PEDAGOGICAL RESOURCES

Further Viewing

Dater, A., & Merton, L. (Directors). (2017). *Burned: Are trees the new coal?* [Film]. Marlboro Productions. https://burnedthemovie.com/
Weldon, J. (Director). (1979). *Log driver's waltz* [Film]. National Film Board of Canada. https://www.nfb.ca/film/log_drivers_waltz/
Wine, S. (Director). (1998). *Fury for the sound: The women at Clayoquot* [Film]. Telltale Productions.

Further Reading

Bowles, P., & Wilson, G.N. (Eds.). (2016). *Resource communities in a globalizing region*. UBC Press.
Braun, B. (2002). *The intemperate rainforest: Nature, culture, and power on Canada's West Coast*. University of Minnesota Press.
Dauvergne, P., & Lister, J. (2011). *Timber*. Polity.
Magnusson, W., & Shaw, K. (Eds.). (2002). *A political space: Reading the global through Clayoquot Sound*. University of Minnesota Press.
Vaillant, J. (2006). *The golden spruce: A true story of myth, madness and greed*. Penguin Random House.

CHAPTER FIVE

From Carbon to Energy

"I just can't stand it anymore," stated Elizabeth May, then-president of the Sierra Club of Canada, as she announced the beginning of what would turn out to be a seventeen-day hunger strike (CBC News, 2001). May's protest was sparked by federal government inaction on relocating residents of Sydney, Nova Scotia, who lived near the infamous Sydney tar ponds. Described as "one of Canada's most contaminated sites" (Venton & Mitchell, 2015, para. 2) or even as "North America's largest toxic waste site" (Weber, 2018, para. 1), the Sydney tar ponds contain, according to various estimates, between 700,000 and more than 1 million tons of contaminated sludge. May, who would go on to become a member of Parliament and leader of the federal Green Party, was already enough of a high-profile public figure that the hunger strike generated considerable media attention. After seventeen days, she agreed to end it when Health Minister Allan Rock announced that the government would fund relocation expenses if tests demonstrated a health risk.

The CBC summarized the findings of a 2001 government report as saying that "the Sydney tar ponds are as safe as any other urban part of the province" (CBC News, 2010, "Dec. 4, 2001" section), but a 2003 study by Health Canada suggested that health risks increased with proximity to the site (CBC News, 2010, "March 26, 2003" section). A $400 million cleanup project, started in 2007, was completed in 2013.

In this story, the organization of coal resources sets in motion a long train of events, with many conflicts and inflection points along the way. Carbon – in the form of coal, in this case – is extracted, mining and industrial operations are developed, rural and urban landscapes are reshaped, and human bodies are affected. As the costs of coal and steel accumulate, other kinds of resources – bodies, money, media attention – are deployed in the struggles over who will bear these burdens.

5.1 INTRODUCTION

Although the word *carbon* has multiple literal and metaphorical uses (Girvan, 2018, pp. 17–23), the focus of this chapter is carbon-based energy sources: fossil fuels. More specifically, we focus here on, and use *carbon* or *carbon resources* to refer to, three types of fossil fuel that have been extracted in Canada: coal, oil (including bitumen), and natural gas. Just as the organization of fish and forest resources profoundly shaped local communities, as discussed in the preceding two chapters, the organization of carbon resources has profoundly shaped contemporary Canadian life. It does so in at least three broad ways. First, as with the organizing of forest resources in the previous chapter, the extraction of fossil fuels entails transforming the local landscape at the site of extraction. Bitumen extraction in northern Alberta (oil sands or tar sands) provides a particularly dramatic example. Figure 5.1, made by renowned photographer Edward Burtynsky, is an aerial image of some of the Athabasca oil sands – described as the largest industrial site in the world – in 2007.

Second, the consumption (burning) of fossil fuels entails the emission of CO_2, a GHG that is an important driver of climate change at a global scale. Thus, fossil fuel consumption entails changes to global landscapes. Third, fossil fuel consumption drives changes to social and urban landscapes, such as the suburbanization (Dale, 1999; Keil, 2017; see also Chapter 7) that comes with car-centred development or the throwaway society that came with the rise of plastics (Monbiot, 2018). It is no exaggeration to say that the modern Western way of life is carbon based.

Unlike the subjects of the previous two chapters, carbon is a nonrenewable resource. Fish populations reproduce over years or decades, and forests reproduce over decades or centuries; it is theoretically

Figure 5.1. "Oil Sands #9," 2007
Source: Photo © Edward Burtynsky, courtesy Nicholas Metivier Gallery, Toronto, https://www.edwardburtynsky.com/projects/photographs/oil.

possible to harvest those resources such that they perpetuate themselves in a sustainable way, hence the ideas of total allowable catch and MSY. In contrast, fossil fuels are formed by dead plants and animals in a process that takes millions of years. Carbon-based fuels, because they take so long to be produced, cannot be sustainably harvested, even in principle. Another way to state this difference between renewable and non-renewable resources is to say that renewable resources are flow resources (i.e., part of something that is endlessly streaming past), whereas non-renewable resources are stock resources (i.e., something of which there is a finite supply). The tension resulting from a non-renewable resource forming the basis for civilization lies at the crux of many of the most profound environmental problems and often expresses itself as a drive to sustain the unsustainable (Bluhdörn, 2007; see also Barry, 2012).

As in the previous chapters, we examine how the specific material nature of carbon-based energy sources affects their ability to be organized and their role as an organizing resource. Each of the three types of fossil fuels under discussion here has a different material nature. Coal is a solid material that is turned into a usable resource by mining. Oil is a liquid, which makes it more difficult to contain but easier to move, and it is transportable through pipelines. Bitumen, found in large quantities in the oil sands of northern Alberta and Saskatchewan, is a heavy form of oil that is so thick that it can only flow through pipelines as diluted bitumen, or "dilbit." Finally, natural gas is a flammable gas that not only needs to be processed to be usable as a fuel but also often needs to be made denser (e.g., liquefied) to be transported effectively. (To be clear about the terminology, the word *gas* as it is commonly used in Canada – short for *gasoline* – refers to a liquid that is derived from oil. *Gas* here, however, refers to a different form of carbon that is normally gaseous, not liquid, in form. In processed forms, it is often used for home heating and cooking – natural gas stoves or furnaces, propane barbecues, etc.)

The development of fossil fuels as a resource can be described as a three-stage history, as social and technological developments made the extraction, distribution, processing, and consumption of different forms of carbon possible: first coal, then oil, then natural gas. Coal, which is the most carbon-intensive form of energy, is increasingly residual in Canada, particularly as coal-fired electricity generation stations are being shuttered as part of efforts to reduce GHG emissions such as CO_2 that cause climate change. Oil is currently the dominant fossil fuel in Canada (and globally). Gas is emerging as an increasingly demanded form of carbon because it is "cleaner" than oil and especially coal, although burning it still releases a significant amount of CO_2.

The transition first to solid (coal) and then to liquid and gaseous (oil and gas) fossil fuel energy sources had social consequences, not only in Canada but globally, that are difficult to overstate. The use of coal power "opened the way to the world's first 'exosomatic' regime: the first to take carbonized solar energy from beneath the crust of the earth and convert it to mechanical energy outside of living bodies" (Fraser & Jaeggi, 2018, p. 97). This story is told in considerable detail at the global level in Timothy Mitchell's (2013) book *Carbon Democracy*. For Mitchell, the specific, material ways that societies get their energy resources has important implications for how politics is structured within those societies: political regimes are deeply connected to dominant energy forms within those societies. Thus, Mitchell (2013) describes contemporary Western societies as "carbon democracies": "Political possibilities were opened up or narrowed down by different ways of organizing the flow and concentration of energy, and these possibilities were enhanced or limited by arrangements of people, finance, expertise and violence that were assembled in relationship to the distribution and control of energy" (p. 8).

In the first instance, because fossil fuels concentrate energy very densely, the advent of fossil fuel combustion meant that large quantities of potential energy could be transported more efficiently. Being able to move fuel relatively easily facilitated both urbanization and large-scale manufacturing. At the same time, as societies became less reliant on forests and pastures for energy sources, this meant that more rural land could be converted to plantation agriculture (Mitchell, 2013, pp. 15–6). Carbon energy thus allowed for the transformation of both urban and rural landscapes, as well as the relations between them.

The concentrated nature of fossil fuels – the fact that coal seams or oil reserves are not as widespread or as evenly distributed as forests – also meant that control over fossil fuels is a potent political resource. Having control of a fossil fuel resource not only means being on the land where (or underneath which) fossil fuels are found, it also requires land ownership, legal (formal) and social (informal) licence to act, technologies, and access to and control over workers. For resource extraction to occur successfully, all of these need to come together, along with the appropriate ideational framework (thinking of coal, oil, and gas as a resource). Conversely, fossil fuel extraction may be thwarted or delayed if any of those components are missing or blocked. In the case studies that follow, we explore the resourcification of carbon and what that process has meant for Canadian society.

5.2 COAL IN NOVA SCOTIA

The sign for the town of Springhill, Nova Scotia, reads "Springhill: You should see us now." Although the intention of the sign is vague, it seems to be alluding to the events that put Springhill on the map: two disasters in the heart of the town's economy – the coal mine – occurring

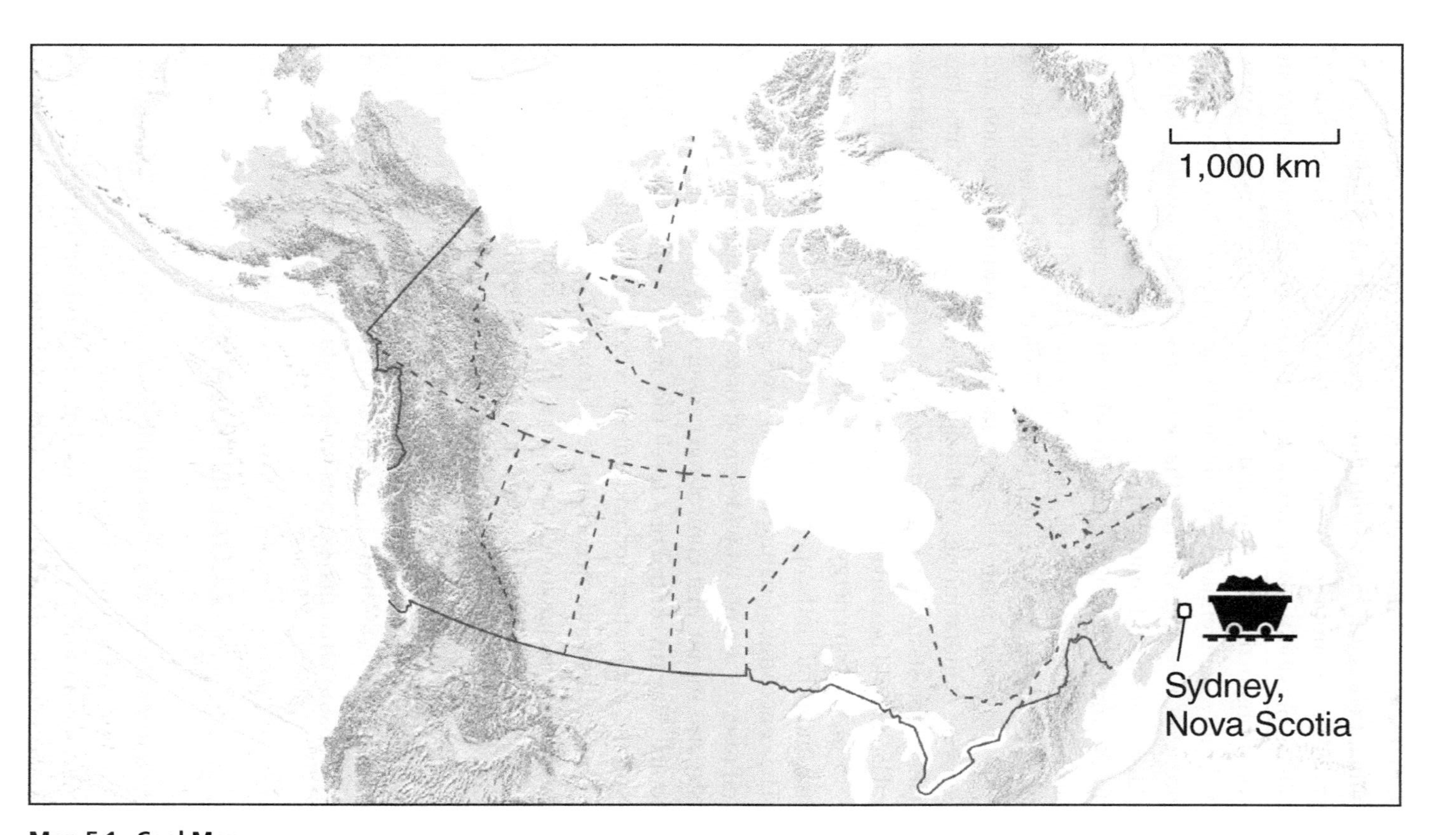

Map 5.1. Coal Map

just two years apart: first, an explosion in the mine in 1956 that killed 12 miners and, second, just two years later, an underground earthquake known as "the bump" that claimed 74 lives in 1958. This latter disaster, especially, captured the imagination of the nation. When 12 survivors were discovered deep inside the coal mine more than five days after the disaster, CBC held its first nationwide live broadcast at the mine's exit while the painstaking work of rescuing the workers took place and they emerged one by one. The mine was closed in the aftermath of the disaster. Myriad cultural moments reference the Springhill mining disaster, not least of which is the 1993 Heritage Minute featuring narration by African–Nova Scotian Maurice Ruddick, who speaks not only about the disaster itself but also about discrimination against him in the aftermath (Historica Canada, 1993). The salience of the moment speaks to issues greater than coal mining in Springhill: it embodies an era in which entire towns lived and died – literally and figuratively – on the coal industry, and it gives voice to the love–hate relationship that many communities had with coal, as well as the extent to which it was the foundation of regional and community identities.

As we have mentioned, fossil fuels are not evenly distributed in the physical landscape. Until the middle of the twentieth century, most coal extraction in Canada happened in Nova Scotia (Map 5.1). Over the course of three centuries, at more than 300 mine sites, about 400 million tons of coal was extracted from Nova Scotia (Department of Natural Resources and Renewables, n.d.).

Mining rights in that province were initially granted to Frederick Duke of York (the second son of King George III). Over the course of the nineteenth century, the industry went through an initial period of growth with many operations being established, followed by a period of consolidation when coal mining operations became more closely integrated with a burgeoning steel industry in Sydney, Nova Scotia: coal as a concentrated energy source was convenient for the high-temperature furnaces needed to forge steel. The prevalence of coal resources in Nova Scotia not only led to industrialization and urban growth in the province, but it also played a role in national development. The Nova Scotia Museum of Industry (n.d.) states that "this coal ... was responsible for the industrialization and urbanization of eastern Canada" (para. 1). If one keeps in mind the significance of building a national railway on which coal-powered trains ran, then in a real sense coal contributed to making possible the colonial settlement of Canada.

Despite the economic and strategic significance of coal resources, and the broader social benefits produced by burning this highly concentrated form of energy, coal miners lived in difficult conditions well into the twentieth century. Although Mitchell's (2013) book mainly focuses on other parts of the world, one Canadian to whom he devotes considerable attention is William Lyon Mackenzie King. King is perhaps best known as one of Canada's longest-serving prime ministers (1921–30 and 1935–48). Before that, however, he served as minister of labour and then as an industrial relations consultant. In a 1918 report, noting what coal miners' unions could do by striking and shutting down mines, King (quoted in Mitchell, 2013, p. 25) said, "Here is power which, once exercised, would paralyze the ... nation more effectively than any blockade in time of war." In other words, King pointed out that access to and control over a labour force was a necessary condition for fossil fuel extraction, and it could not simply be assumed. While King sought to maintain the conditions for the profitable coal extraction that he thought was necessary for national development, coal miners sought to assert some control over coal as a resource for their own ends, including decent pay and safer working conditions. The industry was wracked by a series of 58 strikes in the 1920s, culminating in a general strike in 1925.

Even after the general strike and the (somewhat limited) labour reforms and modernization that followed, coal mining remained demanding and dangerous work. The Nova Scotia Archives (n.d.) accounts 2,426 coal mining fatalities, including more than 500 in the post–World War II period. Most recently, 26 people were killed in an explosion at the Westray mine in 1992. Even this cataloguing of workplace fatalities is the most minimal account of the human toll of organizing coal as a resource, because it does not include, for example, the many premature deaths from mining-induced illnesses.

Over the second half of the twentieth century, the coal mining industry declined as demand decreased and open-pit mines elsewhere, particularly in the United States, were able to deliver coal more cheaply. Following the recommendation of a federal government royal commission (the Donald Commission), the federal government created a Crown corporation, the Cape Breton Development Corporation, or DEVCO, to take over the Dominion Steel and Coal Corporation (DOSCO) coal mines, and the Nova Scotia provincial government created the Sydney Steel Corporation (SYSCO) to take over DOSCO's steel mill. Both DEVCO and SYSCO were wound down around the turn of the millennium. After a period of several years in which there were no regulated underground coal mines operating in the province, the Donkin mine reopened in 2017.

The moving and burning of 400 million tons of coal obviously involves a substantial reshaping of the physical landscape. But it goes beyond that, and beyond the moving of other parts of the earth to get at the coal. The resources that flowed from coal extraction enabled the building of coal mining towns and their associated physical infrastructures, which entailed the movement of other resources to build houses and schools, stores and offices, roads and sewers, and other features of the built environment throughout those towns. It is not much of an exaggeration to say that the integration of coal burning with energy-intensive steel-making made Canada possible. Just as individual coal miners' bodies were sometimes sacrificed in this nation-building process, though, local communities also bore a disproportionate cost of national development. Sydney and its tar ponds, discussed at the beginning of this chapter, provide one of the most dramatic examples.

Perhaps to an even greater extent than with the renewable resources discussed in the previous two chapters, the exploitation of non-renewable coal resources follows a boom-and-bust pattern. Socially valued fossilized carbon (coal as a resource) attracts people to a particular site. Once the resource is no longer worth extracting (perhaps it has been exhausted, or technology has rendered it obsolete, or it is no longer economically viable to extract), it is more difficult for the people there to sustain themselves, even if they have become attached to that location. The legacy of coal mining is thus complicated and extends well beyond the "carbon lock-in" of building physical infrastructure. For example, in 2019, coal burning still accounted for more than 50 per cent of electricity generation in Nova Scotia (down from a high of 80 per cent largely as a result of provincial government policy). Despite the fact that it was dirty, dangerous, and often poorly paying work, the communities produced by coal mining and the identities that those communities fostered, although sometimes romanticized, are for many the subject of deep emotional attachment. This is expressed in a wide variety of cultural forms, including novels, films, and music. The Men of the Deeps, billed as North America's only coal miners' chorus, focus on songs about mining. One of their best-known songs is "Working Man," with the chorus "It's a working man I am / And I've been down underground / And I swear to God if I ever see the sun / or for any length of time / I can hold it in my mind / I never again will go down underground" (MacNeil & The Men of the Deeps, 1988).

5.3 OIL AND BITUMEN IN ALBERTA

Globally, the best-known Canadian carbon resource is almost certainly the oil sands (or tar sands) of northern Alberta (and to a lesser extent Saskatchewan; Map 5.2). We should note at the outset that, in what follows, we generally use the term *oil sands* rather than *tar sands.* As an example of the political power of language, either term is loaded. On the one hand, "'tar sands' is technically inaccurate but favoured by environmentalists as an easy swipe at the gooey mess in which the bitumen petroleum found here is locked up" (Hern & Johal, 2018, p. 72). On the other hand, a Google n-gram search reveals that *tar sands* has been more commonly used than *oil sands* since the 1950s.

Before it was an energy source and a billion-dollar industry, the bitumen found in northern Alberta was used by Indigenous communities as a sealant on canoes (Girvan, 2018). The extensive bitumen deposits were thus well known by the time oil was desired as an energy source, although, as noted earlier, bitumen is a very thick form of oil, which makes it difficult to extract and transport. The ratio of energy needed to energy produced is called the *energy return on energy invested* (EROEI), and the lower the number, the less efficient the process. Compared with conventional oil sources, the EROEI is lower for bitumen. One early proposal to deal with the material properties of bitumen was to detonate a nuclear bomb underground, with the idea that this would liquify the bitumen to the point that it could be extracted more easily (Lamoureux, 2015). Fortunately, this idea was not pursued. Oil sands extraction (first by surface mining, then by other methods to reach more deeply situated bitumen) began in earnest in

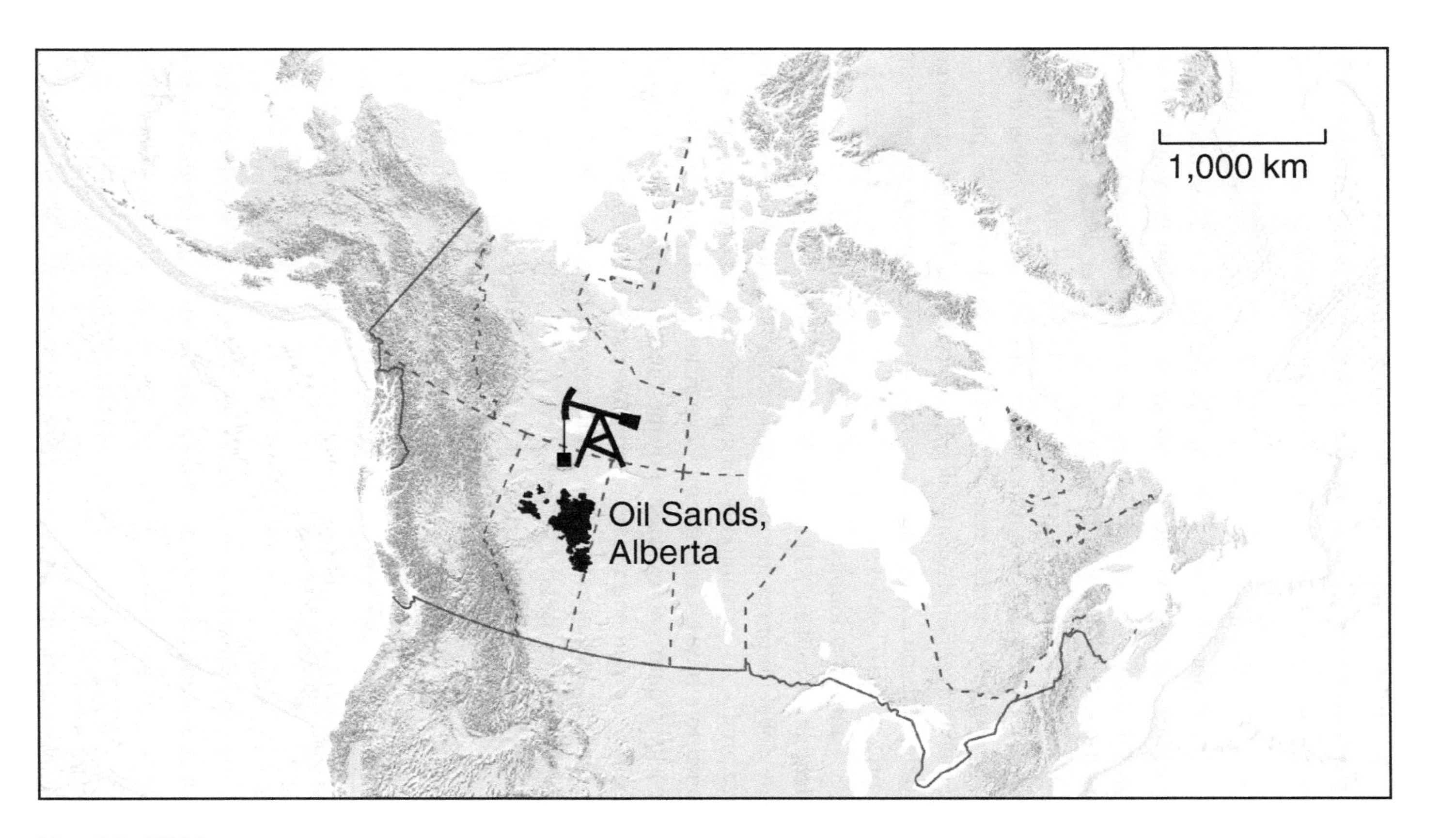

Map 5.2. Oil Map

the late 1960s. Over the past half-century, production has increased to the point where it is now the dominant source of fossil fuel energy in Canada. The oil sands account for about two-thirds of Canada's current oil extraction.

Turning ecosystem components into resources requires a confluence of technological, economic, and political or social conditions. As the nuclear detonation proposal suggests, although the existence of large bitumen deposits was known for a long time, it was not clear how those deposits could be extracted or, in other words, how bitumen, as an ecosystem component, could be turned into a fossil fuel resource. However, it was not just technological change that enabled organizing bitumen as a resource. Because it requires more processing than conventional oil supplies, the viability of oil sands extraction also depends on a particular economic environment: sustained high oil prices. The oil shocks of the 1970s (global oil prices quadrupled in 1973–4 and then doubled again in 1979) made oil sands investments potentially profitable. However, even extraction technologies and high global oil prices are not sufficient for oil sands investors to realize profits. As a commodity, oil must not only be extracted but also moved to a buyer. Consequently, expansion of oil sands production capacity requires an expansion of the pipeline system (either literal pipelines or road and rail networks). We return to the significance of pipeline developments, and the conflicts over them, in our discussion of natural gas in the next section.

The transition from relying on forms of carbon resources that are solid (coal) to those that are liquid (oil) also shifted the balance in terms of the social power of labour, or the extent to which oil could be organized as a resource for workers. In general, being lighter than coal made oil easier to transport, especially by ship, which facilitated international trade and the development of global oil markets. Oil's liquidity also meant that it could be transported by pipeline, requiring less human intervention because it moved by pressure or gravity alone. Compared with coal, the extraction of oil required fewer people, making oil extraction a more capital-intensive (or less labour-intensive) undertaking than coal extraction, thus making higher pay for oil workers a less risky business proposition than higher pay for coal miners. Moreover, workers in the oil industry are working "above ground, closer to the supervision of managers" (Mitchell, 2013, p. 36).

Rather than a solid that must be lifted out by sending people underground and out of sight, oil is a liquid that can be extracted via mechanized wells, with fewer, more closely supervised workers. As Mitchell (2013) summarizes, "These changes in the way forms of fossil energy were extracted, transported and used made energy networks less vulnerable to the political claims of those whose labour kept them running" (pp. 38–9). In other words, whereas Mackenzie King worried about the political power potentially accessible to early twentieth-century coal miners, twenty-first-century oil and gas workers – again, fewer in number, more closely supervised, and more at the mercy of global markets – find it harder to collectively organize carbon resources for their own political purposes.

Adding to Mitchell's (2013) analysis, we note that in the Canadian case, the flow of workers into the oil and gas sector is faster and occurs over longer distances than was the case with coal. Although workers did migrate to coal mining communities, they tended to settle there once they arrived, in contrast to the fly-in, fly-out commuting patterns of many contemporary oil and gas workers. As Hern and Johal (2018) put it (writing specifically about Fort McMurray, in northern Alberta):

> Fort Mac, like most places, is essentially populated by good, decent people just trying to earn a living. It is different from most places, though, in that so many people there are sacrificing so heavily to travel from far away, which changes the calculus.... Not everyone in town has one foot somewhere else, but the dynamic is a foundational one. If it weren't for the spectacular incomes, few people would arrive, or stay. (pp. 81–2)

By 2012, there were 130,000 people working in Alberta who resided in another province, roughly double the number of interprovincial commuters there in 2002 (Alberta, 2017), and often making up large and potentially problematic "shadow populations" for otherwise small communities (Jones & Fionda, n.d.). Ironically, many of these workers travelled to the oil sands from Cape Breton, where communities have remained economically tied to the extraction of carbon, but the crisis of coal has been displaced to Alberta as a spatial fix (see Box 2.3). Indeed, the commuter nature of oil sands labour depends heavily on inexpensive air travel, which is enabled by easy access to cheap oil. In other words, labour depends on oil, and oil depends on labour.

Whereas the nature of coal extraction allowed miners to organize and, at least to some extent, secure a share of economic benefits and a modicum of political power, both the physical nature of liquid fossil fuels and the contemporary political context make oil and gas extraction different. Labour markets and working conditions in the oil and gas sector disorganize workers. Rather than conflicts between workers and owners at the site of extraction, the most contentious political conflicts in the oil and gas sector now tend to happen with its transportation. That is, debates occur over the construction or expansion of pipelines that bring fossil fuels from the site of extraction to market or over the regulation of fracking. Meanwhile, the contemporary political context has changed what these conflicts are about. Coal strikes were largely (although not exclusively; see Malm, 2016, pp. 223–4) about the distribution of benefits from coal mining (including miners' working conditions as well as pay). The fact that coal should be extracted was generally taken for granted. Pipeline and fracking regulation disputes, however, are largely (although again, not exclusively) about whether fossil fuel extraction should be expanded at all, not just about the distribution of benefits from it. This broad shift has less to do with the material nature of oil and gas versus coal and more to do with an increasingly prevalent view that the global costs of expanded extraction (climate change), the local costs of transportation (risks of locally catastrophic spills or leaks), or both far outweigh any potential economic benefits from selling more fossil fuels.

5.4 NATURAL GAS AND FRACKING

Natural gas is often found close to other fossil fuels (oil and coal), with larger proportions of natural gas typically at greater depths from the earth's surface. Although natural gas has been used as an energy source for a long time, its use, and particularly its extraction, has recently become more controversial. As with other fossil fuels, the burning of natural gas produces GHG emissions. Burning natural gas releases less CO_2 than oil or coal, and it is thus sometimes billed as a potential "bridging" energy source in the transition from "dirtier" fossil fuels such as coal and oil to zero-GHG renewables (N. Smith, 2019). However, recent studies have shown that natural gas extraction operations and pipelines leak significantly more methane (which produces significantly stronger greenhouse effects in the short term compared with CO_2; Atherton et al., 2017; for an accessible summary, see Nikiforuk, 2017). More recently, the bridging argument has been seen as dated, given the perceived urgency of transitioning away from fossil fuels altogether, rapidly, to avoid catastrophic climate change.

Apart from its GHG emissions, natural gas extraction has also become controversial as technological developments (as well as higher prices for gas and oil) have made it easier to extract gas from deeper sources. Particularly in the past couple of decades, natural gas fracking has become an increasingly feasible proposition, both technologically and economically. Fracking involves the injection of a fluid at high pressure to fracture (crack open) rock formations in which gas is embedded. One important feature of fracking is the capacity to drill horizontally. In other words, a well at one point on the surface can be used to access gas or oil that is under the surface at some distance away. Although fracking is often associated with natural gas extraction, it can also be used to extract oil that is difficult to access by conventional drilling. Fracking for natural gas is predominant in British Columbia and Alberta, whereas most fracking in Saskatchewan and Manitoba is for "tight oil" (also referred to as "shale oil").

Significant public opposition to fracking exists because of its local environmental impacts, including groundwater pollution and its potential for triggering micro-earthquakes. Along with concerns about methane emissions, fracking opponents also note that the process demands large volumes of water. In 2017, David Minkow reported that Yukon, Quebec, New Brunswick, Nova Scotia, and Newfoundland and Labrador all had moratoria (temporary bans) on fracking. New Brunswick's moratorium was lifted in 2019 (S. Brown, 2019), and there are signs that it is coming back onto the political agenda in Nova Scotia (Baxter, 2019). Meanwhile, in Quebec, "Calgary-based Questerre Energy Corp ... is seeking a court injunction to invalidate the fracking ban" (Van Praet, 2019, para. 9).

As with oil, natural gas pipelines are becoming more contentious as public resistance to expanded fossil fuel extraction increases. According to the Canada Energy Regulator (2021), there are nearly 50,000 kilometres of federally regulated natural gas pipelines in Canada. Both extraction and transportation (pipeline) projects are potential sites of contestation because local communities (whether settler or Indigenous) may be concerned with pollution caused by normal production (extraction) processes or extraordinary spills and leaks. These local conflicts mesh with global conflicts over the scope and pace of fossil fuel extraction and its role in climate change. Of course, communities themselves are also heterogeneous, with community members and even families often split on questions of energy development. Indigenous communities often feature centrally in resistance to fossil fuel extraction and transportation, for three reasons. First, over the centuries-long process of colonization, Indigenous communities have been pushed to the physical margins or hinterlands. As the Canadian and global system of resource extraction continually expands, frontiers are being pushed outward (Coulthard, 2014; Cronon, 1995) and areas that had been marginal become more highly valued. A second is that traditional Indigenous ways of relating to and knowing land are often deeply at odds with the settler colonial view of land as a site for resource extraction. These conflicting views of, and dispositions toward, land are discussed in more detail in Chapter 7. Third, Indigenous Peoples have certain distinctive legal claims – recognized in the Canadian legal system – including the recognition of Indigenous title, and thus they have access to distinct legal resources that can slow or stop applications for fracking permits and pipeline development.

How much control Indigenous Peoples effectively exercise over their traditional territories is a question with a complex and still-evolving answer. Part of the problem, of course, derives from the fact that there are many Indigenous Peoples in Canada, and hence multiple treaties between Indigenous Peoples and the Crown, as well as some Indigenous Peoples (mainly in British Columbia) who live on territory claimed by the Canadian state but not covered by any treaty. The answer to this question has also changed over time, as the Canadian courts interpret Section 35(1) of the 1982 Constitution ("The existing aboriginal and treaty rights of the aboriginal peoples of Canada are hereby recognized and affirmed") in an increasingly expansive manner. Despite this move in a positive direction, at least two concerns persist. One is how far the Supreme Court, which is, after all, still an arm of the Canadian state, will be willing or able to undermine the Canadian state's authority. The second is the limited effectiveness, at least so far, that this kind of appeal to moral authority has had in controlling (including the prevention of) fracking, or more generally in addressing the deplorable living conditions of many Indigenous Peoples in Canada, including some of the environmental (in)justice problems discussed elsewhere in this book.

5.5 CHANNELS IN ACTION: ORGANIZING CARBON

As in previous chapters, we begin our consideration of the channels in action by looking at the role of **governments**. Political scientists often describe countries that are heavily reliant on oil revenues as petro-states. For many of these countries, abundant oil is a resource curse rather than a blessing, because it impedes modernization in a variety of ways (Karl, 1997). Angela Carter (2016) states that "this pattern [of petro-state development] appears to be present in the case of Canada" (p. 296) and that "Canada's dependence on oil is becoming entrenched at the expense of the environment through political and policy choices" (p. 292). Focusing particularly on the majority Conservative government led by Stephen Harper (2011–15), Carter observes four trends in federal government activity that fit with the "oil curse" and show how government choices both reinforce and are reinforced by dependence on a particular resource: "financial support as well as lobbying to defend the industry; hollowing and withholding environmental research; curtailing environmental policies; and stifling criticism of oil development" (p. 296). Although other observers place Canada below the thresholds that are used to define petro-states, there is no question that oil and gas production is a significant component of the Canadian economy, and one that has been increasing rather than decreasing in the twenty-first century.

While the trends seen under the Harper government have arguably abated somewhat under the subsequent Liberal government of Justin Trudeau, they have intensified at the provincial level in Alberta with the election of a majority United Conservative Party government under Jason Kenney in 2019. It is also important to emphasize that the differences between political parties in this case are differences in degree, not differences in kind. Even the federal Liberals under

Justin Trudeau and the Alberta (provincial) New Democratic Party under Rachel Notley (2015–19), as well as earlier governments at both levels, saw oil sands expansion as a desirable path for economic development (MacNeil & Paterson, 2018). Thus, notwithstanding Carter's use of terms such as *policy choices*, we might better focus on the broader structural forces, including an "auto-oil industrial complex" (Freund & Martin, 2000, p. 55) that includes some of the world's largest corporations and is deeply embedded in North American culture (as we discuss later). These forces severely constrain the extent to which these can be described fairly as choices and also point to the importance of understanding channels other than governments. Government decisions in this realm cannot be understood without also understanding the underlying **economies**. For example, recall from Chapter 3 the Harris Report (1990) on the cod collapse, which warned of "irresistible" pressures to overexploit the resource and then to retroactively try to justify that decision. As then-federal Environment Minister Stéphane Dion said in 2005, "There is no minister of the environment on Earth who can stop [the oilsands] from going forward, because there is too much money in it" (as quoted in Haley, 2011, p. 97). This quote suggests that curbing oil sands extraction is unthinkable as a political choice, because money rather than democratic voice drives political decisions. If it is correct, then one can understand Cara Daggett's (2018) claims that anti-democratic "authoritarian politics ... [are] part and parcel of ... a logic of governing that is dependent upon intensive fossil fuel extraction" and that "authoritarianism ... [is at the core] of a contemporary life predicated on burning fossil fuels" (pp. 30–1).

Thus, there is a self-reinforcing quality to the organization of fossil fuel resources in and around Alberta: oil and gas extraction provides obvious economic resources. These economic resources drive development, including attracting workers from other parts of the country. This growth pattern fits Malm's (2016) definition of a *fossil economy* – that is, "an economy of self-sustaining growth predicated on the growing consumption of fossil fuels, and therefore generating a sustained growth in emissions of carbon dioxide" (p. 11).

Yet this type of economy is not sustainable over the longer term, because fossil (carbon) fuels are a finite and effectively non-renewable resource and because the atmosphere's capacity to absorb CO_2

without causing dangerous warming (climate change) is also limited. Some two centuries after the beginning of the Industrial Revolution, the world is fast approaching, if not already at or beyond, that limit, and there is an imperative to "return to the flow" of renewable, non-GHG-emitting sources of energy, even though the obstacles to such a transition are daunting (Malm, 2016, pp. 367–88).

Let us return to the broader structural forces, and particularly to the realms of **culture and ideas** as channels for the organization of fossil fuel resources. As noted earlier, we can say without exaggeration that Canada has or is part of a carbon-based civilization. The Petrocultures Research Group (2016) uses the term *petroculture* to describe these carbon-based civilizations that are

> shaped by oil in physical and material ways, from the automobiles and highways we use to the plastics that fill up every space of our daily lives. Even more significantly, fossil fuels have also shaped our values, practices, habits, beliefs and feelings ... [such as] the ideas and ideals of autonomy and mobility that have become essential values to people around the world. In a very real way, these values are fueled by fossil fuels, as are so many of the other values and aspirations that we have come to associate with the freedoms and capacities of modern life. (pp. 9–10)

For a concrete example of the petroculture in action through everyday culture, look at Petro-Canada's (2019) "Live by the Leaf" ad campaign, which associates an oil and gas corporation with national identity and culture. The ad's tagline, "We share more than a country, we share a way to live" is an excellent illustration of petroculture. Unlike the Molson ad discussed in Chapter 1, which strongly identifies Canada with wilderness, this ad combines images of wilderness and urban settings (the ad begins with a shot of a large apartment building) and includes more than a dozen distinct shots of fossil fuel–powered vehicles (cars, trucks, boats) in its sixty seconds.

In Chapter 2, we discussed staples, with a relatively narrow focus on their production (extraction). One can also think of staples as objects of consumption. As we noted in Chapter 2, Marshall McLuhan (1964) described staples as commodities that are widely used and part of consumers' everyday lives. Whereas economists such as Harold

Innis (1930) and Mel Watkins (1963) were interested in how staples production shapes or skews economic development, McLuhan (1964) was more interested in how staples consumption shapes cultural life: "Cotton and oil ... become 'fixed charges' on the entire psychic life of the community. And this pervasive fact creates the unique cultural flavor of any society" (p. 35). Whether or not people as individuals are involved in the extraction of fossil fuels, it is difficult, if not impossible, to imagine the practices of everyday life (culture) for most Canadians – from Zambonis at the local rink to drive-through coffee shops – outside of a regime of fossil fuel consumption.

The normalization of fossil fuel consumption is also reflected in and reinforced by the everyday language that people use. This extends well beyond the description of the zone of extractive development in northern Alberta as the oil sands or tar sands. As Matthew Hoffmann (2019) points out, a car that is plugged in is routinely described as an electric car and a car that runs on both gasoline and battery power is a hybrid car. But a car that runs on gasoline only is simply a car. And if one asks how far it is between Toronto and Ottawa, the common answer is five hours – that is, the driving time – rather than the number of kilometres (450, for the record).

At the same time, and as Hoffmann (2019) admits, the evolution into something other than a petroculture will require more than just a change in the words that people use. Hern and Johal (2018) make the important point that the realm of culture and ideas is intimately tied to the flow of material resources: "Fort McMurray is the last place Suncor or Shell need to win hearts and minds – that battle is settled every day by filling wallets" (p. 76). Petroculture is built on a petro-economy and petro-state.

Canada is also populated by petro-communities and petro-bodies. Like fishing and lumber towns, **communities** that are sites of fossil fuel extraction (and processing) are built and rebuilt around the needs of that extractive industry. In the case of carbon, we could point to the boom, first, of coal mining towns across Cape Breton and the northern part of mainland Nova Scotia and, second, to the industrialization boom in Sydney, Nova Scotia. The explosive growth of Fort McMurray is a more recent dramatic example. As Hern and Johal (2018) describe, "Bitumen ... built this place, which continues to add residential subdivisions and strip malls. The population – 2,000 in 1967 when oil sands extraction became commercially viable – is now 30 times as large" (p. 74). More generally, as outlined in the introduction to this chapter, fossil fuels make possible certain kinds of **built environments** (Mitchell, 2013). For instance, suburbs rely on cheap transportation, dense cities rely on food imported from outside the city's boundaries, and airports exist because of relatively cheap fuel.

At the same time, one of the striking features of Canadian cities, which experienced more of their growth after the advent of fossil fuel energy than many cities in Europe and Asia, is their sprawling character, made possible by the ability to travel significant distances by car relatively cheaply. Another feature of fossil fuel energy, particularly when combined with the technology of the internal combustion engine, is that it allows people to be much more mobile. Indeed, Matthew Paterson (2007) argues that "contemporary societies can be defined as *dromocratic*: ruled by movement and acceleration" (p. 5) and that "automobility" is "the dominant form of daily movement over much of the planet (dominating even those who do not move by car)" (p. 132). These are particularly striking developments when one considers that cars have been mass produced for only barely a century and only became a widely owned consumer good since the end of World War II. The emphasis on speed and acceleration, iconically represented in fossil-fuelled vehicles, runs so deep that societal evolution or development is imagined largely as progress, which literally means "forward motion." As Paterson succinctly puts it, "To *be* modern is to *be* mobile" (p. 121). This orientation toward mobility means both that built environments are designed to facilitate movement (and particularly car-based movement; see Chapter 7) and that motion and dynamism is taken as a sign of progress: in a world dominated by networks and flows, built environments are considered to be thriving to the extent that they are always newly rebuilt environments (Cannavò, 2007).

If access to cheap and abundant fossil fuels has shaped communities, how then have they shaped and organized human **bodies and identities**? One obvious way is the bodily toll that is exacted from fossil fuel extraction workers and people who live close to or downstream from large-scale extraction sites. Examples range from black lung disease among coal miners to higher incidences of various forms of cancer in communities (and particularly Indigenous communities) downstream from the oil sands, such as in Fort Chipewyan (Edwards,

2014; Eggerston, 2009). Similar costs are borne by those close to points of burning or consumption of fossil fuel, from mercury and fine particulate emissions in coal-fired electricity-generating stations to ground-level ozone from gasoline-burning car engines.

There are other, more subtle ways that people's bodies are enabled as well as constrained, or empowered as well as harmed or disfigured, by the organization of carbon resources. The idea that people are "petro-subjects" helps to capture the idea that they do not so much individually choose to consume carbon resources but rather that they are born into a society that expects them to and is organized so that people have little choice but to consume carbon resources. For example, most roads are set up to accommodate cars and buses rather than bikes and pedestrians; solar panels are a "green addition" to be paid for rather than being the default option; shipping is cheapest by cargo vessel and not by sailing ship; and real estate prices incentivize commuting (Malm, 2016, p. 12).

It is thus not surprising that there are also bodily benefits to living in a fossil-fuelled, high-energy society. On the one hand, fossil fuels allow people to move their bodies faster and farther than they could in the past. People can consume foods and other goods from farther afield; goods purchased online can be delivered to one's door quickly thanks to cheap and abundant fossil fuels, and when people go to the grocery store, they are presented with bounty from around the world: grapefruits from California, mangoes from the Philippines, tea from China, quinoa from Peru, coffee from Guatemala, olive oil from Spain, and so on. Contrast this to a century ago, when one of the authors' Canadian grandmother received her annual orange at Christmastime and her first taste of pineapple occurred when she was well into her forties. On the other hand, a UK study found that a dominant response to the perceived (and real) danger posed by cars was to reduce children's mobility and independence (Hillman et al., 1990). Clearly, patterns of consumption have changed dramatically in response to the rapidity and affordability of long-distance transportation: the bodies, behaviours, and expectations of petro-subjects are shaped and reshaped by the organization of carbon resources and production of car-centred communities.

A final, less tangible, but no less real set of impacts has to do with the affective or emotional response to, or psychological disposition of, living in a carbon-based civilization. We have already mentioned the valorization of mobility and speed, which is one dimension of this. And one should not underestimate the pleasures that can be derived from putting carbon to work, from year-round exotic produce to winter getaways in tropical locations. More extreme and aggressive versions of these pleasures can be found in what Daggett (2018) describes as "petro-masculinity": a particular form of masculine gender identity that is performed in and through practices of "fossil fuel extraction and consumption" (p. 44). At the same time, the climate change produced by the excessive burning of fossil fuels also produces psychological anxiety or distress, variously identified as climate grief, ecological grief, or "solastalgia" (Albrecht et al., 2007; Cunsolo & Ellis, 2018).

5.6 SUMMARY AND CONCLUSIONS

As we wrote this book, debates about carbon taxes or other forms of carbon pricing were in the news. Governments ensuring that the price paid for carbon resources reflects the actual cost (including environmental damage) of carbon consumption is important. However, carbon taxes and other forms of carbon pricing, such as cap-and-trade schemes (carbon markets) or the purchase of carbon offsets, intervene primarily at the final stage of carbon extraction: consumption. Such approaches presume, rather than challenge, the idea of carbon as a resource.

Although carbon is not a living ecosystem component like fish or trees, it nevertheless occupies ecosystemic space and can be considered an ecosystem component. Moreover, fossil fuels were at one time in the distant past living ecosystem components. In this sense, Kathryn Yusoff (2013) provocatively suggests that people can even reverse their way of thinking about where agency lies in the human–fossil fuel relationship: fossil fuels make contemporary human life possible, and thus "humanity built on the dead matter of the Carboniferous is not just underpinned by that materialism but is an expression of it" (p. 784).

Even if one does not take it that far, we have shown in this chapter how carbon energy has had a profound effect in organizing modern Western (and global) society more generally and collective settler colonial life in Canada in particular. The specific material nature of various carbon-based fossil fuels, as well as their geographic location, has enabled Canada's development in particular ways both obvious – such as the rapid growth of Fort McMurray – and subtle – such as imported foods and expanded suburbs. The burst of energy from Nova Scotian coal helped to make Canada a transcontinental European settler state, and oil and gas resources have shaped or skewed the country's political, economic, and spatial (physical) development over the past half-century. Beyond their material manifestations, fossil fuels have also embedded themselves as ideas, organizing how people think about their identities, both collectively and individually.

DISCUSSION QUESTIONS

1 In 2005, then-Environment Minister Stéphane Dion said, "There is no minister of the environment on Earth who can stop [the oil sands] from going forward, because there is too much money in it." Was he right? What are the implications of Dion's claim?

2 The resourcification of ecosystem components depends on their usefulness to humans. In what ways is carbon useful to communities in Canada? How have we organized this resource, and how has it organized us?

3 In many ways, those who benefit from burning carbon for energy are spared much of its costs. In what way might the introduction of carbon taxes change the ways in which carbon resources are organized?

PEDAGOGICAL RESOURCES

Further Viewing and Listening

Historica Canada. (1993). *Heritage Minutes: Maurice Ruddick* [Video]. https://www.historicacanada.ca/content/heritage-minutes/maurice-ruddick

Kheraj, S. (Host). (2018, September 27). *Carbon democracy and Canadian history* [Video podcast]. Network in Canadian History and Environment. https://niche-canada.org/2018/09/27/natures-past-episode-62-carbon-democracy-and-canadian-history/
Petro-Canada. (n.d.). *Live by the leaf.* https://www.livebytheleaf.ca/
Petro-Canada. (2019). *Petro-Canada live by the leaf* [Video playlist]. YouTube. Retrieved from February 21, 2023 from https://www.youtube.com/playlist?list=PLWMRezgdS5q7rbElgAOw924-EFunE9xsF
Petrocultures Research Group. (n.d.). *About petrocultures.* https://petrocultures.com/

Further Reading

Carter, A.V. (2020). *Fossilized: Environmental policy in Canada's petro-provinces.* UBC Press.
Howe, M. (2015). *Debriefing Elsipogtog: Anatomy of a struggle.* Fernwood. [On fracking in New Brunswick]
Kingsolver, B. (2013). *Flight behavior: A novel.* HarperCollins.
Paterson, M. (2007). *Automobile politics: Ecology and cultural political economy.* Cambridge University Press.

CHAPTER SIX

From H_2O to Water

The November 28, 2005 cover of *Maclean's* magazine showed then-President George W. Bush menacingly drinking a glass of water next to a threatening headline ("America Is Thirsty"; *Maclean's*, 2005). The cover image tapped into a visceral reaction among Canadians opposed to the idea of exporting water south of the border. What triggers this common, horrified reaction? Is it objectionable because it would put a price on what Canadians see as their right? (Spoiler: there is nothing about a right to water in the Canadian Constitution.) Is it because water is somehow central to settler Canadian identity, and selling it is antithetical to Canada's very identity as a country? Is it the prospect of water being a trigger for conflict with our closest neighbour? The answer is probably a little bit of each of these, and in this chapter we explore why that might be the case. Unlike refined petroleum or national parks, water has always been necessary for human survival, which is perhaps why its resourcification is troubling, although that is also true of food, and the same emotions are not triggered by discussion of selling tomatoes or canola oil. Or maybe it has more to do with identity – that is, the colonial Canada based on a certain idea of wilderness featuring an overturned canoe on a dock next to a misty lake.

6.1 INTRODUCTION

Water is not like the other ecosystem components discussed in previous chapters. Not only has it become a resource that is extracted in and of itself as a commodity, but it also allows many other resources to be extracted: it takes water to grow crops, to mine, and to turn trees into pulp and paper. Indeed, it is arguably these enabling qualities that have facilitated water's resourcification. For example, water's utility for irrigating crops, producing hydropower, maintaining healthy fish populations (i.e., "hosting" a healthy fishery), transporting goods (i.e., facilitating the transportation of forestry, mining, and other products), absorbing and transporting waste, and extracting and processing oil and metals make it useful not only in and of itself, but also as a substance that allows for industrial-scale extraction and processing of other resources.

Water itself is, of course, essential for human and ecological health. It is very much an ecosystem component: water is ingested and excreted by people, plants, and non-human animals and is constantly circulating around the planet. The shift to conceptualizing water as a resource, however, is something altogether different and involves the leveraging of water's various properties to facilitate the large-scale production, extraction, and processing of commercial goods. Jamie Linton (2010) explains that people's conception of water as H_2O ("modern water") is unique in considering water as homogeneous and "not complicated by ecological, cultural, or social factors" (p. 8). Perhaps unsurprisingly, this shift from ecosystem component to resource is part and parcel of the development of the capital-intensive industries foundational to colonial Canada's nation-building endeavours. Consider, for example, the "Log Drivers' Waltz" discussed in Chapter 4. Although this culturally iconic video is ostensibly about the forestry industry, log drivers only exist because of the rivers that allow for the easy transportation of logs from forestry sites to shipping ports.

Indeed, water's importance in the colonial Canadian imaginary cannot be understated. Any talk of water exports raises the ire of Canadians determined to protect "their" water, and the prospect of increased water utility rates draws cries of "rights to water" despite the fact that no such rights exist in Canadian law. Although Canada's "myth of abundance" may indeed exaggerate the abundance and availability of fresh water in Canada, it is also true that Canada's relatively water-rich geography has organized Canadians in particular ways. Many of Canada's earliest settlements were determined by access to fresh water for drinking and transportation – Winnipeg, at the confluence of the Red and Assiniboine Rivers, is a classic example – and many resource communities were settled subsequent to the development of the appropriate (water-related) extractive technologies, for example, the colonial settlement of the Alberta prairies alongside the development of irrigation districts.

In this chapter, we trace some of the ways in which water has been used in Canada's nation-building project – diversions and damming and drinking water provision – before exploring the six channels in action.

6.2 DIVERSIONS AND DAMMING

Diversions and dams are both about controlling large volumes of water. Diversions move water from one place to another to meet demands most often related to municipal needs or drinking water, and dams control the flow of water – first stopping it, and then directing it through turbines to generate electricity. In this sense, dams and diversions are both classic examples of how humans have organized nature. They are also examples of how ecosystem components are transformed into resources for economic and societal benefit. Less obviously, they show how water has organized people. For example, cities can be powered by dams hundreds or even thousands of kilometres away; billions of taxpayer dollars support the construction of dams; and international treaties have been crafted to address diversion. Moreover, dams present a colonial–Indigenous binary in high relief, because it is often Indigenous lands that are flooded in dam construction. Because dams and diversions require extensive expertise, significant amounts of capital, and large tracts of land, all require some degree of state involvement. At the low end, this involvement could include approval for building and administration of environmental impact assessments; at the high end, it can involve funding dam construction, partnerships with industry to fund and construct the dam, and connecting dams to existing state-run electrical distribution systems. The examples in this chapter illustrate how Canada is involved in transforming the landscape and moving people in the name of controlling and managing water as a resource and how water has, in turn, organized Canadian social, political, and economic landscapes.

6.2.1 Diversion

Water diversion is the moving of water from one area to another. Typically, diversions are built so that water can be used for irrigation (the biggest driver of diversion worldwide), for hydropower, for flood control, and for municipal water supply.

Being able to move large quantities of water and control its quality was a significant driver of urbanization, particularly starting in the nineteenth century (Bakker, 2010). The history of water provision in Winnipeg provides a good example of how the organization of water resources intersects with discourses of economic development and settler colonialism (Map 6.1). The aqueduct built in the early twentieth century that sought to turn Winnipeg into the "Chicago of the North" involved expropriating land from the Shoal Lake 40 First Nation's reserve and effectively turning that community into an artificial island, making it more difficult for them to access clean drinking water (Perry, 2016). The community was under a boil-water advisory for a quarter-century, from 1997 until 2021. After intense political activism and media attention in the mid-2010s, an all-season road providing access to the community was built in 2018–19, and a water treatment facility was subsequently built.

Diversion takes work. If you have hiked with a day's worth of water in your pack, you know that water is heavy: Recall that one cubic metre of water (1,000 litres) weighs one ton. The average Canadian household uses 328 litres per person per day (Environment Canada, 2011), meaning that Canadians on average use one ton of water every three days. Because that water is heavy, it takes significant infrastructure and energy to move it from a source to a treatment plant, to the place where it will be used, and then from that place to a wastewater treatment facility. The farther the water must travel,

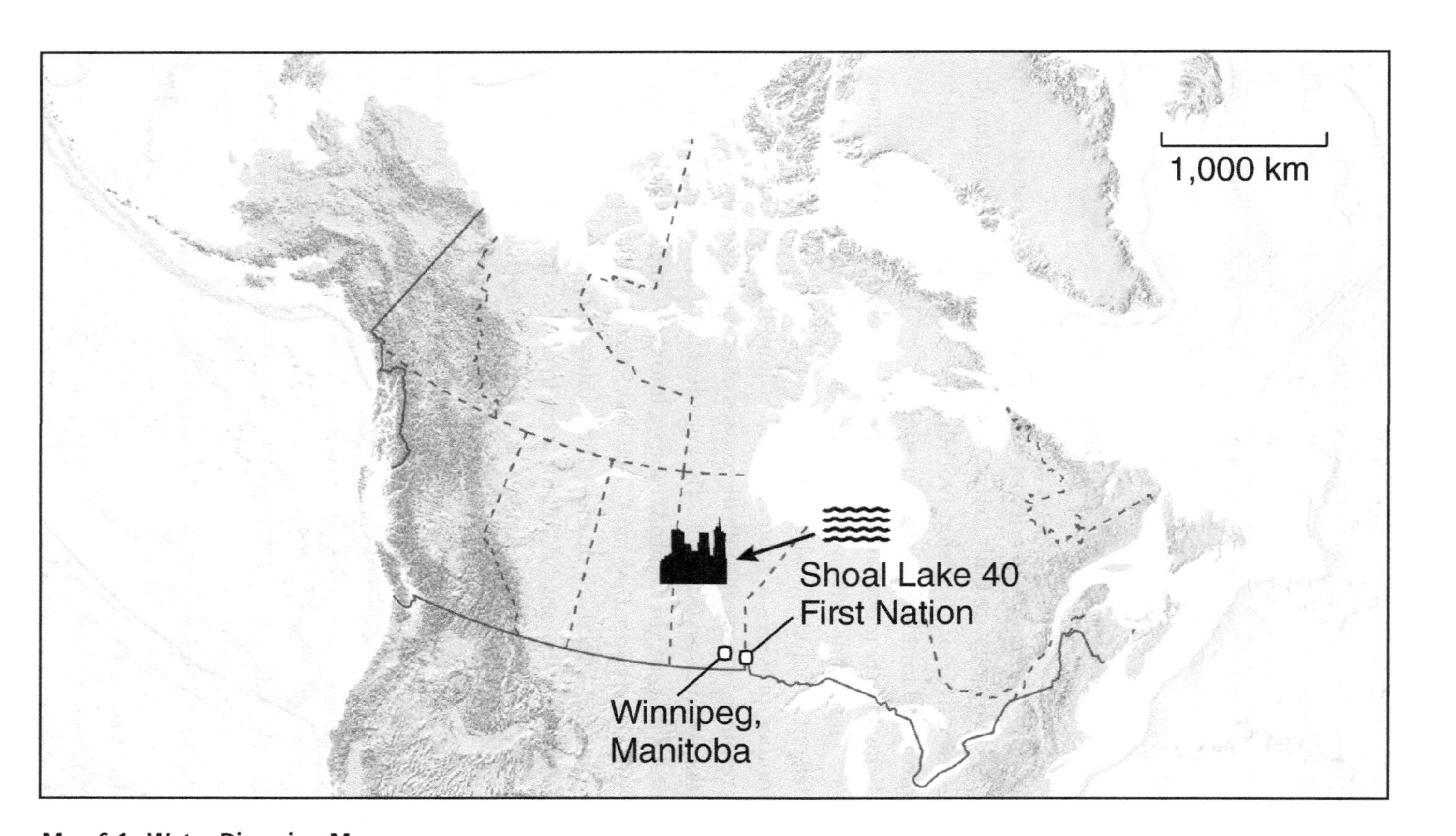

Map 6.1. Water Diversion Map

the more complicated and expensive the diversion, and the greater the potential environmental impacts. Indeed, even when technically and financially possible, diversions raise questions about how much water can be moved without incurring significant environmental damage. Moreover, diversions proposing to move water from one jurisdiction to another are politically problematic. Here is an example: the Great Lakes Compact is a 2008 agreement between all the states and provinces that border the Great Lakes to prevent water diversions outside of the Great Lakes basin. Waukesha, Wisconsin, lies just outside the basin and needs an additional water supply because its groundwater sources are contaminated with radium. Under a 2016–17 agreement, water will be diverted from Milwaukee (within the basin) to Waukesha if the same amount of water is treated and returned to Lake Michigan (the source of Milwaukee's water; Riccioli, 2017). In effect, the agreement artificially expands the Great Lakes basin by incorporating Waukesha within it. News of this agreement was met with controversy, especially from Canadians worried that allowing a water transfer out of the basin would open the floodgates to draining the Great Lakes more generally. Water – and, specifically, large bodies of fresh water – has somehow become enmeshed with a version of Canadian identity that is fiercely opposed to water exports, however perceived (Biro, 2002; for examples, see Holm, 1988, and Council of Canadians, 2020).

6.2.2 Damming

Dams work by blocking the flow of a river, creating a reservoir of water behind the blockage. Water is then moved through the dam in a controlled way; as it moves through the dam, it turns turbines that generate electricity. The flow of water can be increased or decreased to meet energy demands, and the resulting electricity can be transported over long distances to serve energy needs far away from the dam. Dam building is controversial because of the dam's effects on the river and its people: the dam itself impedes the travel of fish and other marine life, and the reservoir of stagnant water behind the dam floods the surrounding area, often causing the loss of culturally significant areas and property; flooded areas also contribute to climate change when the underlying vegetation decomposes and releases methane

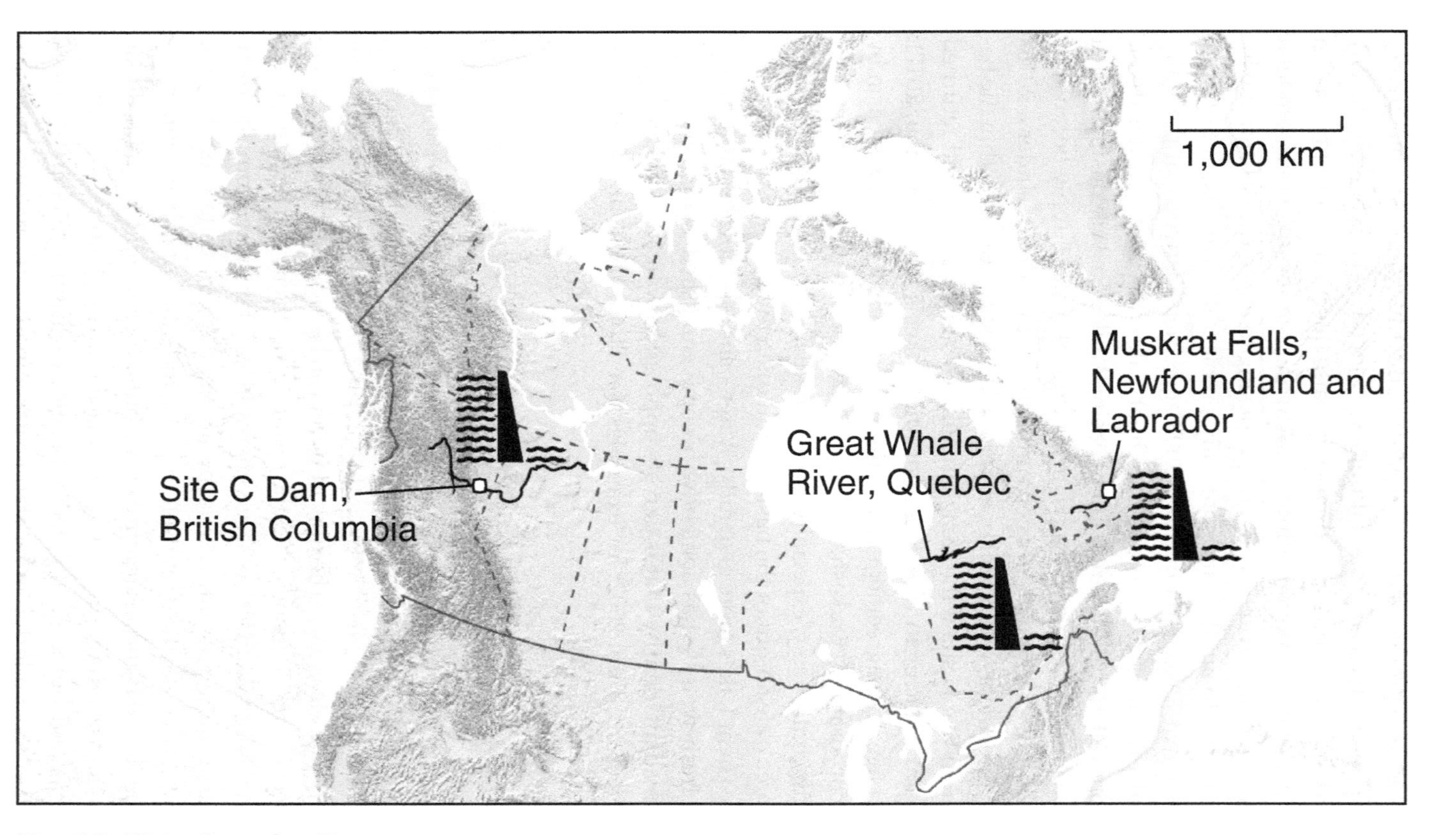

Map 6.2. Water Damming Map

and carbon dioxide. Although the benefits of dams can accrue to people far away, the costs – including displacement – are borne by communities proximate to the dam site.

Large dams are both physical and symbolic: they display the state's capacity to reshape territory at an enormous scale – "pushing rivers around," in Ken Conca's (2005) phrase. They were often seen as emblematic development projects, signalling the arrival of modernity (World Commission on Dams [WCD], 2000, p. xxix). When working at that scale, and particularly when backed by the ideology of development, there was often little regard for more localized consequences for human and ecological communities (B. Forest & Forest, 2012). Local human communities did, however, often resist such disruptions to the landscape. In the second half of the twentieth century, those local resistance movements increasingly communicated with each other and other allies, creating globally coordinated resistance that was able to push back against dam proponents: states, large infrastructure companies, and international financial institutions such as the World Bank. By the 1970s, global construction of large dams – defined as any dam 15 metres or higher – peaked globally. In the late 1990s, in response to the increasingly problematic nature of large dam construction, the WCD was created to study the issue. The WCD's final report, *Dams and Development: A New Framework for Decision-Making*, published in 2000, concluded that although dams can produce considerable benefits, those benefits have not been distributed equitably. Going further, the report concluded: "In too many cases an unacceptable and often unnecessary price has been paid to secure those benefits, especially in social and environmental terms, by people displaced, by communities downstream, by taxpayers and by the natural environment" (WCD, 2000, p. xxviii).

Although the WCD stopped short of recommending a complete ban on large dam construction, it did raise the bar on how such projects should be undertaken (WCD, 2000, p. iii). Of course, many large hydroelectric dams had already been built in Canada in the late nineteenth and early twentieth centuries. According to the International Commission on Large Dams (n.d.), Canada ranks seventh in the world in the number of large dams, with 1,169. In four provinces with abundant river flows (British Columbia, Quebec, Manitoba, and Newfoundland and Labrador), 90 per cent or more of electricity is generated from hydroelectric dams.

Although hydroelectric power is seen as a green energy source because it does not involve burning fossil fuels or the extraction of non-renewable resources, building dams involves making significant landscape transformations. Large quantities of material must be transported to the site to block the natural flow of the river. Once the river is successfully dammed, a reservoir of water is created behind the dam, flooding land that was previously above the water line. The weight of water in large reservoirs is sometimes enough to affect tectonic plates and cause earthquakes, a phenomenon known as reservoir-induced seismicity, and the methane emitted from rotting plant life in the flooded reservoirs is a contributor to global climate change.

The Great Whale project is an example of how humans have organized water in Canada (Map 6.2). The Great Whale project proposed damming the Great Whale River (*Rivière de la Baleine*) as part of the second phase of Hydro-Québec's massive James Bay Project. Construction of the dam would have generated electricity for sale in New York State. Like the first phase, this second phase was opposed by local Cree and Inuit Indigenous Peoples. The first phase in the 1970s eventually led to a negotiated settlement (the 1975 James Bay Northern Quebec Agreement, which was the first comprehensive land claim agreement in Canada; for more information, see the link in the Pedagogical Resources section at the end of this chapter) that saw the project proceed with some modifications, but opposition to the second phase was stauncher, partly because by that time the negative environmental effects of the first phase had become visible. The second phase of the project was ultimately suspended because local Indigenous groups were able to effectively mobilize opposition by "jumping scales" (Cox, 1996; Herod & Wright, 2002): a canoe trip by Cree leaders and allies from Hudson Bay to the Hudson River (in New York) brought attention to the issue and (among other factors) helped convince the State of New York to withdraw from an agreement to buy power from this project (CBC Radio, 2015; Figure 6.1).

The journey took six weeks. As one of the paddlers, Matthew Mukash, explained,

> The Odeyak was put onto a stage at Times Square and it became the centre of attention for the media. And our leadership was

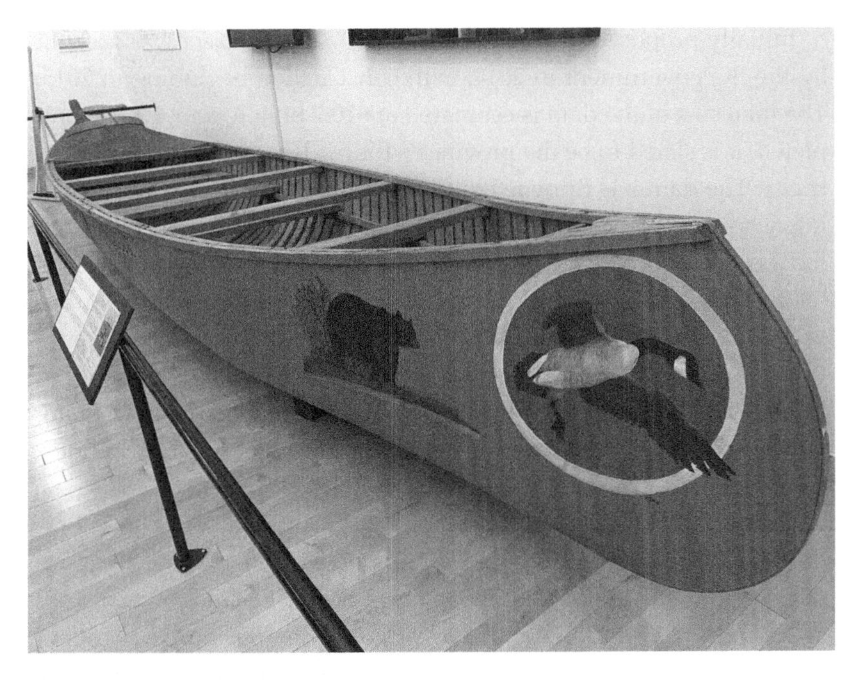

Figure 6.1. The Odeyak
Source: Photo by Kory Saganash. Published with the permission of Aanischaaukamikw Cree Cultural Institute.

> given the opportunity to speak before a crowd of about 10,000 people at Times Square, so that part was very emotional for all of us.... Of course it was a big undertaking because here you had a deal that was signed by the State of New York and Quebec that was about $17 billion that you wanted to kill. The strategy was to kill the market for electricity in the United States because that's where Hydro-Québec wanted to sell the power. (CBC Radio, 2015, para. 5)

Two more recent examples of controversial dam projects are the Site C dam on the Peace River in British Columbia and Muskrat Falls in Labrador. Both show how the "organizing" of water, in this case in the form of dam building, is controversial for both environmental and legal reasons, and yet it continues because of other ways in which people are organized: Canadian life is dependent on cheap electricity, and Canadian economies are dependent on resource exports.

Initially proposed in the 1970s, the Site C dam project was greenlit by the BC government in 2014, with construction beginning in 2015. The total cost of the dam is estimated at $10.7 billion, and, when completed, it is slated to be the province's fourth-largest producer of electricity. The dam has drawn significant criticism for its environmental impacts – namely, its impact on the Peace River – and their effects on Indigenous Peoples in the area. BC Premier John Horgan expressed his frustration with a project his government did not initiate or approve: Horgan said of the dam that "this is not a project that we favoured, it's not a project that we would have started" (as quoted in Morgan, 2017, para. 2) and that proceeding with the dam is "'making the best of a bad situation' because the cost to cancel the under-construction power project would be $4 billion" (para. 7).

The Muskrat Falls project on the Lower Churchill River in Labrador is a dam to produce hydroelectric power to be exported via undersea cable (the Maritime Link) to Nova Scotia. Because of its history of coal mining (see Chapter 5), Nova Scotia remains more dependent than other provinces on coal-fired electricity generation. Importing hydro-generated electricity is part of Nova Scotia Power's strategy to meet provincial government targets to reduce GHG emissions. Like the many dams that came before them, both projects face resistance, including direct action protests, from both local settler communities and First Nations.

6.3 DRINKING WATER

Drinking water is another example of how the Canadian state has resourcified water. In this case, "organizing" has meant drawing a line between water in general and drinking water, when, of course, the same water keeps circulating around the biosphere, changing forms and locations as it goes (the water in your water bottle could be recirculated dinosaur urine!). In this model, drinking water is governed by health departments and is protected, treated, distributed, and assigned a financial value accordingly; non–drinking water is a different category altogether and is governed primarily through environment departments. Drinking water also shows how water as a resource has organized people: water is notoriously difficult to

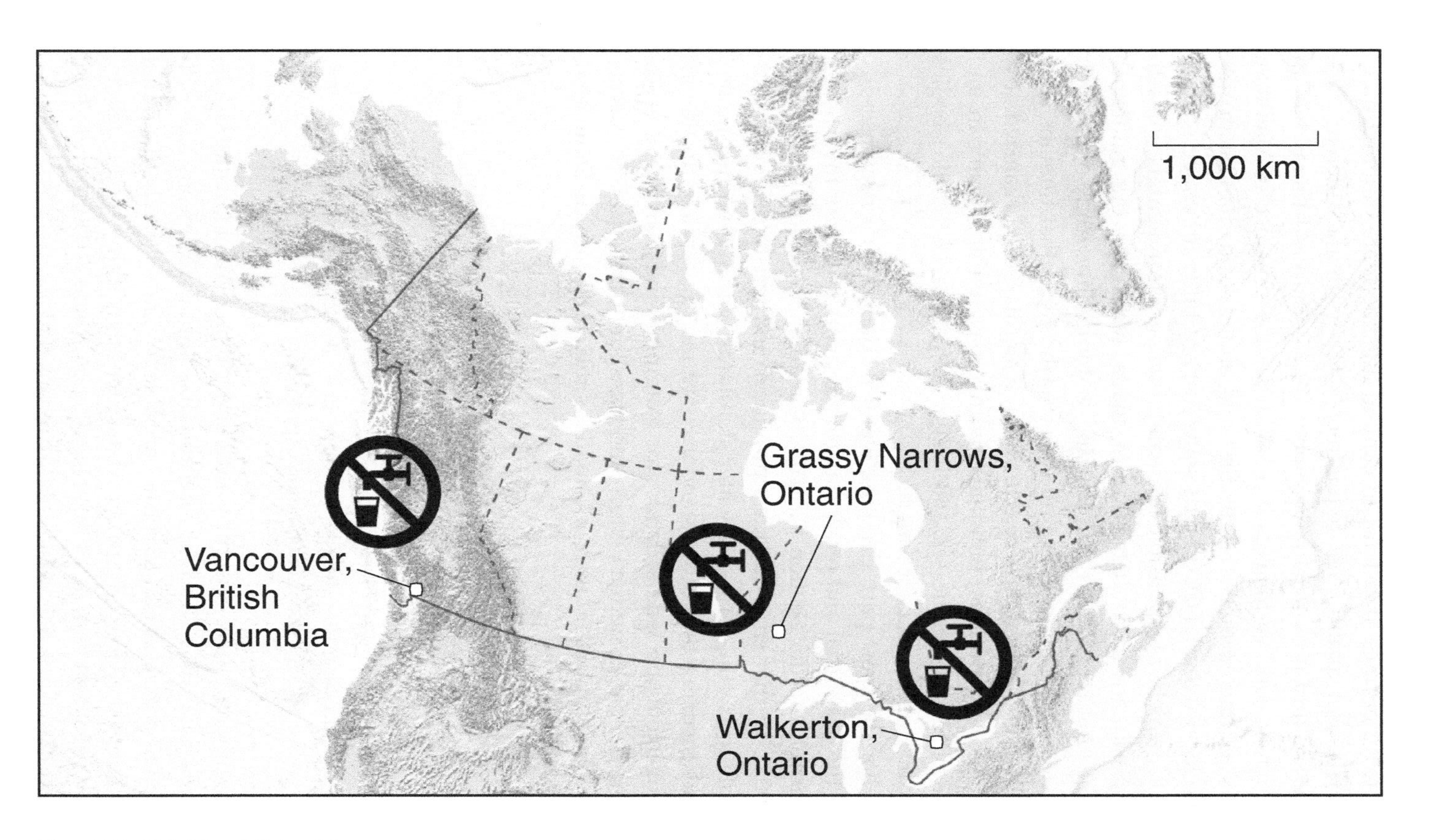

Map 6.3. Drinking Water Map

govern, because it is hard to contain and there are different qualitative and quantitative requirements for its many different uses. Federal and provincial governments have organized themselves in different ways to manage it. For a time, the province of Manitoba had a Ministry of Water: the first and only in Canadian history. At the federal level, ECCC holds some responsibility for water; so does FOC, which is responsible for maintaining environmental quality in fish habitat; so too does Health Canada, which sets out guidelines (note: not regulations) for drinking water quality. And so although people are organized around water in the sense that humans have settled in places with access to water, water has also organized governance institutions, buildings (with pipes running water in and out of homes, businesses, and industries), and even people's bodies (which are 70 per cent water).

For the most part, drinking water in Canada is a responsibility of the provincial and territorial governments as part of their health portfolio under Section 92 of the Canadian Constitution (Leeson, 2002). To that end, Canada's provinces and territories are responsible for setting drinking water standards within their jurisdiction and ensuring that those standards are met. Exceptions to this arrangement are federal lands, for example, First Nations reserves, national parks, military bases, and transboundary areas. The decentralized approach taken in Canada is relatively rare; as Dunn et al. (2014) note,

> Canada is the only Group of Eight (G8) country and (along with Australia) only one of two Organization for Economic Cooperation and Development [*sic*] (OECD) member states that does not have legally enforceable drinking water quality standards at the national level – despite "best practice" World Health Organization (WHO) recommendations. (p. 4635)

Although there exist the Canadian Drinking Water Quality Guidelines (CDWQGs), Canada's decentralized approach means that "Canadian provinces and territories are able to choose: (i) whether or not to adopt the CDWQG; (ii) to what extent they apply the guidelines, and (iii) whether to make them legally enforceable" (Dunn et al., 2014, p. 4637). In practice, this flexibility means that

> Eight of the 13 jurisdictions have legally binding drinking water regulations.... However, only one of these eight jurisdictions, the Northwest Territories, has adopted all 94 CDWQG into provincial regulation. Nova Scotia and Manitoba have adopted the majority of the CDWQG into provincial regulation. The remaining five jurisdictions (Alberta, Saskatchewan, Ontario, Québec and Yukon) have partially adopted the CDWQG into legislation. Whilst BC follows the CDWQG in their entirety ... only the microbial indicators are legally binding. According to the British Columbia Water and Wastewater Association ... BC is currently the only provincial or state level of government in North America that does not explicitly identify drinking water standards beyond bacterial indicators. Non-legally binding approaches are also taken in New Brunswick, Prince Edward Island, Newfoundland and Labrador and Nunavut. (Dunn et al., 2014, pp. 4637–8)

With such a patchwork of standards and enforceability, it is perhaps unsurprising that Canada has a high number of boil-water advisories. Watertoday.ca (see Pedagogical Resources) maintains a live map of advisories across the country, and on any given day the number is likely to be well above 500. Indeed, in 2015, an article in the *Canadian Medical Association Journal* reported that there were 1,838 active advisories (Eggertson, 2015). Of those, 136 advisories were in First Nations communities, and that number does not include First Nations communities in British Columbia, where the overall number of water advisories is highest, at 544. Moreover, many of the advisories in First Nations communities are long-standing, such as Shoal Lake 40 First Nation, mentioned earlier. In the community of Kitigan Zibi in Quebec, for example, "unacceptable levels of uranium are responsible for a do-not-consume order that has been in place since 1999" (Eggertson, 2015, p. 488), an order that remained in place as recently as 2021 (Ireton, 2021).

There is no single cause of poor drinking water quality in First Nations communities, and it cannot be explained by small populations and remoteness alone. Indeed, a 2012 study reviewing the capacity of small, remote Ontario communities to provide safe drinking water concluded that the significant discrepancies between small and remote First Nations communities and small and remote non–First Nations communities suggests that "the problem with drinking water in First Nations is possibly more than just a financial or technological problem" (Walters et al., 2012, p. 21). Contributing to the problem, then, is the constitutionally ambiguous responsibility for drinking water provision in Indigenous communities. Health is a provincial responsibility, but First Nations reserves are federal, leaving many communities in limbo while federal and provincial governments admonish each other for inaction on the First Nations water file. This finger-pointing problem was described by Kathryn Harrison (1996) in her book *Passing the Buck* as a "hot potato," wherein each level of government is hesitant to act because the financial and technical costs are high, and the political gains are low. On a hopeful note, she points out that political gains can be high when public attention is drawn to an issue, suggesting that widespread public pressure to address the unacceptably high number of water advisories in Indigenous communities is perhaps one way to address the issue.

A handful of case studies illustrate some of the ways that drinking water protection (or the absence thereof) has played out in practice (Map 6.3).

6.3.1 Vancouver, 2006

In November 2006, heavy rain rinsed silt and mud into the drinking water reservoirs in the Seymour and Capilano watersheds on Vancouver's north shore. Drinking water in Vancouver (and most everywhere else) is treated with chlorine, which is not effective when water is highly turbid – that is, cloudy – as shown in Figure 6.2.

In this instance, the issue was not that the water had become contaminated with bacteria, but rather that the city's treatment system was no longer effective, so Vancouver residents were advised to boil water at home to kill any potential bacteria. Of course, boiling turbid water does not make it clear – just bacteria free – so demand for bottled water was high through the duration of the advisory. After twelve days, when the turbidity cleared, the boil-water advisory was lifted.

6.3.2 Walkerton, Ontario, 2000

In May 2000, six people died and thousands were sickened after drinking tap water contaminated with *Escherichia coli* bacteria. The

Figure 6.2. Turbid Water
Source: The Canadian Press/Chuck Stoody.

tragedy was the result of a perfect storm of factors: intense spring rainfall rinsed cow manure from a cattle-breeding operation into the soil and groundwater near an improperly sealed well. When contaminated water from that well reached the municipal drinking water treatment facility, it should have been treated with chlorine. Instead, the drinking water plant operators – brothers Stan and Frank Koebel – were not using chlorine because they and other Walkerton residents preferred the taste of "raw" water (Parr, 2005). Even this lack of chlorination would not have been as catastrophic if the Koebel brothers had had the training to understand the lab results from routine testing; instead of immediately issuing a boil-water advisory when tests showed the presence of *E. coli*, they submitted bottled water samples. As Scott Prudham (2004) notes, at the inquiry into the tragedy, "Incredibly, *Koebel testified to the public inquiry that he had never read the province's guidelines on unsafe drinking water, and did not know what E. coli were*" (p. 350). A final contributor was that these samples were processed not at a provincial lab as had been the case for decades, but, as a result of recent provincial government budget cuts and efforts to trim red tape, at a private lab that had no contractual obligation to report health concerns to the provincial health department (Prudham, 2004). As lawyers Theresa McClenaghan and Richard D. Lindgren (2018) put it, "Red tape reduction was not a side issue that might have contributed a little bit to the catastrophe. It was the central factor" (para. 4). It was this toxic combination of weather, faulty infrastructure, public pressure, operator incompetence, and environmental deregulation that proved fatal.

After the tragedy, the province called for a public inquiry. The inquiry was led by Justice Dennis O'Connor, who produced a two-part report totalling more than a thousand pages: the first part of the report documents the events that contributed to the tragedy (D.R. O'Connor, 2002a), and the second part sets out a strategy and recommendations for avoiding such tragedies in the future (D.R. O'Connor, 2002b). The Walkerton tragedy was a wake-up call not only across Ontario but in communities across the country; many provinces adopted Justice O'Connor's recommendations in the hopes of preventing a similar tragedy.

6.3.3 Asubpeeschoseewagong Netum Anishinabek–Grassy Narrows, Ontario, 1962–?

Known in Ojibwe as Asubpeeschoseewagong Netum Anishinabek and in English as Grassy Narrows, this community provides a different example of contamination resulting from industrial activity that affects an Indigenous community. The main contaminant in this case is mercury, which was a by-product of the pulp and paper processing facility near the community. The effects of mercury poisoning are primarily neurological: symptoms include things such as muscle weakness, tremors, and lack of coordination; in children it causes cognitive delays such as problems with fine motor skills and speech. Mercury is also passed from mothers to babies during pregnancy and through breast milk. So, even if there were a way to immediately eliminate all the mercury in the water and fish, mercury poisoning would persist in the community for generations.

The community of Asubpeeschoseewagong Netum Anishinabek–Grassy Narrows is no stranger to struggle and upheaval. In the 1950s,

Ontario Hydro flooded much of the territory to produce "cheap and clean" power; in the 1960s the community was relocated to its current location 120 kilometres east of Winnipeg. Then, between 1962 and 1970, Reed Paper dumped more than 50,000 pounds of mercury into the English–Wabigoon River system (Ilyniak, 2014; L. Simpson et al., 2009), and the "Anishinabek people, who relied on the water from the river for drinking and fishing, were not told about the mercury until several years later. For over a decade they continued to drink the contaminated water and to eat the contaminated fish" (L. Simpson et al., 2009, p. 8). Natalia Ilyniak (2014) describes the effects of mercury as follows:

> Mercury is highly poisonous to humans, with frightening health implications (Vecsey 1987).... When humans ingest mercury, "[they] absorb 95–100% of it into the body, excreting it at a very slow rate" (Vecsey 1987: 294). Symptoms include: speech, taste, and smell impairment, difficulty swallowing, choking, blurry vision, loss of strength (CBC 2012), "tunnel vision, loss of co-ordination, numbness, tremors, loss of balance" (CBC 2009, July 29), and accelerated body deterioration, leading to impaired motor functioning (da Silva in Ball 2012). Toxins travel to large organs and in pregnant women the mercury settles in the fetus, leading to birth defects (Vecsey 1987). There are also "hidden effects, including things like miscarriage and lowered resistance to alcohol and infections" (ibid: 295). These are not just issues of the past. (p. 45)

The mercury problem cannot be solved by boiling the water: that would just lead to boiled mercury. The only solution is to stop drinking the water and eating the fish. This solution, however, is especially problematic because of historical context. Situated in Treaty 3 Territory, the Anishinaabe People of Grassy Narrows were dependent on fish for subsistence, for culture, and for income (many worked as commercial fishers and in the lucrative fishing tourism industry). As a result, mercury contamination of the water and fish was a three-part disaster: community members' bodies were poisoned with mercury, they lost their jobs as commercial fishers and tourist fishing guides (Kerr, 2010, as cited in Ilyniak, 2014), and they were effectively cut off from a key dimension of cultural identity. As Michif scholar Max Liboiron (2021) notes in their discussion of a similar case of polychlorinated biphenyls in the Mohawk community of Akwesasne, "The stakes of not eating the [contaminated] food are cultural genocide, where the languages, practices, knowledge, and thus relations with Land are killed" (p. 107). Second, as Ilyniak (2014) points out, the tragedy exemplifies environmental injustice both in the causes of the tragedy and in its solutions. Namely, the dumping of mercury was possible because Reed Paper was permitted to operate in traditional territory, and the solution – that is, to stop drinking the water and eating the fish – had disproportionate effects on Indigenous Peoples. In addition to the devastating health impacts of mercury poisoning, the sudden loss of fish as a primary source of food and employment means that

> Food is now purchased in Kenora [the closest town]. Small budgets lead to imbalanced diets contributing to type 2 diabetes, cancer, thyroid disease, and plasto-illnesses never seen in the community previously (p.o. 13 Nov 2012). The economic losses from the fishing ban sent the community from a "95% employment to 95% unemployment" rate (Vecsey 1987: 294). Unemployment led to high rates of alcoholism, personal withdrawal, negative self-evaluation, and routine violence (Clement 2003). Today the community struggles to be dry and participate in traditional activities like trapping, rice harvesting, and hunting as much as possible (p.o. 28 Sept 2012). At the same time, major corporations reaped the benefits. Between when the mercury was discovered in 1970 and 1984, Reed's operations (who changed their name to avoid liabilities, but remained under the same management) received net profits of $235 million. (Ilyniak, 2014, pp. 53–4)

6.3.4 Drinking Water: Summary

The Vancouver, Walkerton, and Asubpeeschoseewagong Netum Anishinabek–Grassy Narrows cases underscore some important themes in Canada's drinking water landscape. First, they show how fragmented Canadian water governance frameworks are. Each of these crises had a different cause and a different solution, addressed by a different level of government (municipal, provincial, and Indigenous, respectively). There is no legally binding set of national drinking water standards, nor is there a nationally consistent protocol for addressing crises when they do arise. Second, they show how diverse the factors affecting water can be. Indeed, a range of factors – natural events, managerial failures, budget cuts, colonialism, corporate influence, and history – can trigger and exacerbate crises. Third, they highlight the variety of possible contaminants: turbidity, *E. coli*, and mercury are highlighted here but by no means is this a complete list. Given Canada's enormous landscape, multiple levels of government, and economics, it is perhaps unsurprising that diversity is the thread tying together drinking water crises. This diversity is explored in the next section, in which we explore how Canada's geographic and cultural diversity has enabled fragmented governance, while, at the same time, Canada's fragmented water governance landscape has intensified geographic, historical, and racial inequalities.

6.4 CHANNELS IN ACTION: ORGANIZING WATER

How **governments** engage in the formal governance of Canada's water can be described as a patchwork: provinces are responsible for drinking water, waste water, and water licensing for industrial activity (including mining) and agriculture, and the federal government is responsible for transboundary waters; fish habitat protection; and water in federal territory, such as national parks, military bases, and First Nations reserves. In addition, there is a patchwork within each level of government: at the federal level, for example, there are 20 federal agencies and departments with responsibility for some dimension of water governance. This polycentric model extends even to some pieces of legislation. For example, the federal *Fisheries Act* is administered jointly by FOC and ECCC. This patchwork is well documented elsewhere (Bakker, 2006; Dunn et al., 2014; Weibust, 2009) and serves as the backdrop to this section rather than its focus. Here, instead, we focus on two effects of institutions germane to this book. First, we explore the ways in which Canada's institutions have facilitated the establishment of what Donald Worster (1985) calls a "hydraulic society," and second, we discuss how Canadian water institutions embody the problematic nature–society binary that this book aims to challenge.

In his foundational book *Rivers of Empire*, Donald Worster (1985) frames the American West as a hydraulic society – a society with a complex social division of labour driven by a desire to increase agricultural productivity through increasingly sophisticated irrigation technologies – that is, a society in which whoever controls water controls society. The equating of control over water with control over society is particularly true when control of water is a capital-intensive venture requiring extensive changes to the **built environment**: dams, pipes, pumps, filters, irrigation channels, and other technologies so costly that the only viable owners are either the state or a small number of large companies. In these circumstances, posits Worster, society pivots around control of, and access to, water, although this power structure is not necessarily always evident.

Worster (1985) was writing about the American West, but many of his insights can be translated to a Canadian context because many areas of this country have similarly only been able to be permanently settled with large populations as a result of significant water infrastructure. Examples here include dams in Quebec and British Columbia and irrigation technologies in the arid agricultural west. In both of these cases, the biophysical reality of place combines with technological advancement and a state desire for large-scale settlement to create exactly the kinds of power structures described by Worster.

The relationships between Canadian governmental and non-governmental institutions and the country's evolution as a hydraulic society are many. For example, many of Canada's dams are owned by provincial Crown corporations – that is, companies owned by rovincial or territorial government – such as Manitoba Hydro, Hydro-Québec BC Hydro, NB Power, Newfoundland and Labrador Hydro, and the Yukon Energy Corporation. In these cases, provincial governments have invested resources in constructing dams to provide energy (and in some cases, flood control and water for irrigation) for their citizens. Similarly, controlling irrigation water in arid regions allows for permanent settlement of sparsely populated areas ("go west, young man!"). As Alberta Water (the institution responsible for water in that province) writes in a description of its history:

> In much of southern Alberta, there is not enough rainfall and moisture to naturally sustain agricultural crops. However, there is abundant sunshine and heat that can contribute to growing

> many different crops if water were not a limiting factor. Early in the settlement of Alberta, it was recognized that agriculture would not be successful in the southern region without an abundant and assured supply of water to irrigate fields. Irrigation Districts were organized and granted water licenses to divert large quantities of water from the tributaries of the South Saskatchewan River, primarily the Oldman (St. Mary, Waterton and Belly) and Bow Rivers. (Alberta Water Portal, 2012, "Water Used for Irrigation" section)

This example speaks to some of the confluences among geography, institutions, infrastructure, and colonialism: state interest in settling the west (regardless of the reality that it had been inhabited by Indigenous Peoples for thousands of years) was facilitated by the state control of water, which enabled large-scale economic activity in an otherwise arid area. Certain kinds of settlements (e.g., farming **communities**) were facilitated through physical alterations to a landscape that, other than aridity, had growing conditions conducive to permanent agricultural settlement and, by extension, economic activity. At the same time, these landscape alterations could only be imagined as possible to the extent that people subscribed to the tenet of human separation from, and mastery over, nature. Government documents encouraging and planning for both intensive and expansive permanent settlement in the west provide ample evidence of faith in this separation of people (settlers) from nature (B. Forest and Forest, 2012; Martin, 2009; Perry, 2016). The people–place binary is also replicated, facilitated, and enabled through Canadian water governance institutions more generally. A classic example is the separation between drinking water and what we call "other water." In the wake of the Walkerton tragedy, one of Justice O'Connor's recommendations was a multi-barrier protection system that started with the protection of drinking water sources. The rationale in this case was that preventing contaminants from entering bodies of water that feed into public drinking water systems would reduce the overall risk of drinking water contamination in the future. In implementing this recommendation, nineteen Source Water Protection Areas were created across the province, each charged with developing a Source Water Protection Plan to reduce the risks to the drinking water sources in their particular area.

For the purposes of this book, the key point about Ontario's *Safe Drinking Water Act, 2002*, and the resultant Source Water Protection Plans, is that water in the province was separated into two categories: drinking water (i.e., water for human use) and other water (i.e., environmental water). Note that because water supplies flow into people's homes through a single pipe, drinking water encompasses all household uses of water, not just drinking. This separation perpetuates the nature–society binary by protecting only that water that is destined for human consumption and, relatedly, by creating a legislative and policy structure that replicates and promotes the idea that water for people is somehow separate from water that is not directly headed for human consumption. Of course, basic hydrology asserts that water is constantly circulating through human and environmental systems and that the distinction between source water and other water is fleeting, evaporating (so to speak) through the various stages of the water cycle.

As much as water does flow around the planet and is managed at different scales, though, for most Canadians, it is municipally scaled water utilities that are responsible for delivering water to their household. Most of these corporations are publicly owned, although some public–private partnerships exist, whereby municipal governments contract with private companies for some or all aspects of water provision (Aït-Ouyahia, 2006; Siemiatycki, 2016). Canada's municipally driven system gives rise to yet another patchwork with different mixes of public- and private-sector involvement and different plans for ensuring financial and hydrological sustainability. Some municipalities may charge a flat rate for connection to the system regardless of how much water is used, and others impose volumetric charges (households that use more water pay more). More important, whether the water utility is publicly or privately owned and whether costs are imposed on a fee-for-service or volumetric basis, the water that flows through taps is imagined as a commodity, and households are framed as customers.

This last point shows the significance of **culture and ideas**: what do people take water to be? Globally, the idea that water should be treated as a right rather than as a commodity has been widely debated (Mirosa & Harris, 2012), and in 2010 the UN General Assembly formally "recognized the right to safe and clean drinking water and sanitation as a human right" (Sultana & Loftus, 2015, p. 98). However, as legal scholar David Boyd (2011) notes, in Canada the right to water is not guaranteed in law:

> The Canadian Government does not recognize the right to water, either internationally or domestically. When the United Nations (UN) General Assembly approved a resolution recognizing water as a human right in 2010, 124 countries supported the resolution while none were opposed. Canada was among forty-two countries that abstained from voting, and it has a history of blocking international efforts to recognize the right to water. The Canadian Constitution Act, 1982 (Constitution) does not explicitly acknowledge a right to water. There is no federal legislation explicitly recognizing the right to water in Canada. To date, no Canadian court has acknowledged the right. (p. 85)

Moreover, because municipalities are "creatures of the province" (not recognized constitutionally as having a distinct set of rights), they have limited abilities to raise funds for their operations, including water provision. Most urban drinking water infrastructure is hidden from view (much of it literally buried underground), and most household water consumption is done in reflexive, almost unconscious daily practices (Sofoulis, 2005). As a result, absent a crisis such as the one in Walkerton mentioned earlier, prioritizing water infrastructure upgrading or even maintenance may require intensive political education campaigns. Estimates of the infrastructure deficit in Canada are as high as $250 billion (Water Canada, 2016).

Indeed, money and water become inextricably linked when water becomes a resource, stressing the significance of **economies**. Perhaps the most obvious form of water commodification is bottled water. Some bottled water is sourced from municipal water corporations (effectively, bottled tap water, possibly further processed in some way before bottling). In other cases, water bottling companies pay fees (often relatively small) to provincial governments for a licence to extract large volumes of water from a natural source and then bottle it for sale at a considerable markup. For example, in August 2017, the Ontario government increased its fee from $3.71 per million litres to $503.71 per million litres – a substantial increase, but still only a tiny fraction of a cent per bottle (CBC News, 2017). Although much attention is focused on consumers' decisions to pay a premium price for bottled water (e.g., Biro, 2019; Brei, 2018), questions can also be raised about the decision to turn that water into a resource (commodity) in the first place. In one high-profile case, Nestlé purchased a well in Elora, Ontario, for its water bottling operations, outbidding the Township of Centre Wellington, which wanted to secure a water source for municipal expansion (Bueckert, 2016). Should the licence to tap a water source be left to the highest bidder? If not, who should get to make those decisions, and on what basis?

As the drinking water cases suggest, the organization (and unequal distribution) of water resources does not happen only through market mechanisms, but turning water into a resource often means calculating the costs and benefits of physically organizing (moving, storing, treating) it in monetary terms. Turning water into part of the economy is both made possible by and reinforces a view of the world around one as resources rather than ecosystem components.

Of course, there are perspectives on water other than economic ones. Boyd (2003) writes that "from Inuit kayaks, Aboriginal canoes, and the voyageurs, to Group of Seven paintings, novels by Michael Ondaatje, Margaret Atwood, and David Adams Richards, and summers in cottage country, water is at the heart of Canada" (p. 13). There is little doubt that Boyd's observation here is correct, that water has a distinctly cultural value that cannot be easily reduced to a price tag. These icons feature prominently in Canadian (popular) culture, with a resulting myth of abundance (Bakker, 2006): the idea that Canadians are particularly blessed with plentiful water supplies and that therefore watery experiences rightly figure into Canadian identity and experience.

The effects of this myth of abundance can be seen in the fact that Canadians are among the highest users of water in the world (OECD, n.d.). Some of this consumption happens at an individual or household level, but it is also the case that this sky-high consumption is a function of an economy structured around cheap water that allows for the proliferation of water-intensive activities such as irrigated agriculture and water-intensive mining. Although individuals and households might be persuaded to reduce their water use through voluntary appeals to change daily behaviour, other drivers of high water consumption are not, because the myth of abundance is built into an interlocking set of legislation, policy, planning, and permitting. The **identity** channel comes into play here, too. Water being central to Canadian identity is a double-edged sword. On one hand, Canadians' pride in having "the most water in the world" may make it easier to generate support among Canadians for water protection or conservation issues. On the other hand, water issues may be assessed using inappropriately nationalist frames. Somewhat like fish in Chapter 3, water resources may be conceptually constructed as national resources ("Canadian water") even though water is a flow resource that often crosses national (and other human-made) borders. For example, water naturally flows around and through an undetectable national border in the Great Lakes. Although having "their" water taken by the United States has been a perennial source of anxiety for Canadian politicians, authors, and activists since at least the 1970s, P. Forest (2010) finds a dozen transboundary local water supply agreements, in which communities on either side of the Canada–United States border share a water supply.

As with other resources, Canadian attitudes toward and interactions with water suggest a profound ambivalence: a valuation of natural environments and experiences and at the same time a reliance on resource extraction. Canadians' cultural attachment to water – as seen, for example, in beer ads and the *Maclean's* magazine cover discussed at the outset of this chapter – may lead to a belief that Canadians by their nature are water protectors or use water sustainably. This can make it more difficult to address water issues that challenge this nationalist frame. At the same time, however, the Canadian state and society have maintained a long-standing support for resource extraction industries (particularly mining of various kinds) that entail high levels of local water pollution and, in the case of fossil fuel extraction (see also Chapter 5), contribute to global climate change with its attendant impacts on water resources. Although many Canadians can take abundant water supplies for granted, a focus on the recurrent problems with drinking water in First Nations communities might lead to the conclusion that an inability or unwillingness to provide secure access to clean water is at the heart of the Canadian state.

6.5 SUMMARY AND CONCLUSIONS

Throughout this chapter, our focus has been on how water has been abstracted both materially and conceptually and the effects of this abstraction. Water diversion and damming projects, in Canada as in the rest of the world, are among the most significant projects of landscape transformation that human societies undertake. They have immediate and dramatic impacts on the non-human environment. However, they also produce long-term and equally dramatic transformations in human communities. Canada's population today is more than 37 million and more than 80 per cent urban. By comparison, at the time of Confederation in 1867, Canada's population was less than one-tenth of what it is today and was less than 20 per cent urban.

The physical movement of water that made these transformations possible was itself made possible by the conceptualization of water as a resource: "modern water" (Linton, 2010). No less than with fish, forests, or carbon, a particular way of thinking is required to conceive of water as a resource capable of being organized. Its extraction and movement from one part of the environment to another is enabled by conceiving of it in abstract terms, quantifiable in litres, cubic metres, or acre-feet, and as inert and subject to physical and chemical manipulation to produce drinking water or water that is suitable for other specific purposes.

Although Canadians have organized water, water resources have in turn organized Canadians. Canada's hydraulic society has perpetuated a myth of abundance and with it a culture that sees resource-intensive and high-impact lifestyles as an aspirational ideal. Particularly with respect to water resources, a significant component of this ideal involves a dualist separation of urban and industrialized waterscapes on the one hand from natural waterscapes (cottage country, camping, etc.) on the other, with the latter serving as an idealized site for escape from the former.

Finally, following Worster (1985), a hydraulic society is not only one in which water is intensively managed as a natural resource. It is also one in which the expertise required for intensive resource management produces and reinforces significant inequalities. The benefits of a hydraulic society are far from equally distributed. If the state's provision of access to clean water is, as Bakker (2010) suggests, a material emblem of citizenship, it is an emblem that is unevenly distributed, particularly (but not only) between Indigenous Peoples and settler Canadians, as demonstrated through differentiated access to safe drinking water across the country.

Thus, water is simultaneously organizer and organized – and both processes are enabled by Canadians' conceptualization of water as a resource rather than as a component of the environment.

DISCUSSION QUESTIONS

1. The previous chapter framed Canada as a petro-state. This chapter describes Canada as a hydraulic society. What do these two descriptors have in common, and how are they different? Can they both be true? How might your answer differ regionally or over time?
2. How might the themes of this book – commodification, dispossession, and the perpetuation of the nature–society binary – contribute to the disproportionate number of boil-water advisories in Indigenous communities?
3. In what ways is water similar to or different from the other resources described in previous chapters? How do these similarities and differences affect the transition from H_2O to water?

PEDAGOGICAL RESOURCES

Further Viewing

Baichwal, J., & Burtynsky, E. (Directors). (2012). *Watermark* [Film]. Mercury Films. [Canadian documentary about global water use].

Daniel, I., & Page, E. (Directors). (2019). *There's something in the water* [Film]. Giant Pictures. [Documentary about environmental justice struggles in Nova Scotia].

Watertoday.ca. (n.d.). Boil water advisories. http://www.watertoday.ca/map-graphic.asp

Further Reading

Clancy, P. (2014). *Freshwater politics in Canada*. University of Toronto Press.

Maich, S. (2005, November 28). America is thirsty. *Maclean's*, pp. 26–32. https://web.archive.org/web/20201203123812/https://archive.macleans.ca/article/2005/11/28/america-is-thirsty

Makivik. (n.d.). *JBNQA* [James Bay Northern Quebec Agreement]. https://www.makivik.org/corporate/history/jbnqa/

Schmidt, J. (2017). *Water: Abundance, scarcity, and security in the age of humanity*. NYU Press.

CHAPTER SEVEN

From Land to Property

Figure 7.1 is an iconic image of a standoff. It shows two men staring at each other, their faces only a few centimetres apart. On the left, wearing a helmet and military garb, is Canadian Forces soldier Private Patrick Cloutier. On the right, wearing camo-patterned fatigues, including a bandana that covers most of his face, is Mohawk warrior Brad Larocque. The picture was taken on September 1, 1990, almost two months into a land dispute that captured national attention. The town of Oka, just north of Montreal, planned to expand a golf course onto land that included a Mohawk burial ground. Although the Mohawks had sought consistently since the eighteenth century to assert their rights to that land, the British, and then later Canadian, state repeatedly denied their claims. When the golf course expansion began, a group of Mohawks blockaded road access and refused to obey a court order to dismantle the blockade. On July 11, Sûreté du Québec (SQ; the provincial police force) tried unsuccessfully to enforce the order, and SQ Corporal Marcel Lemay was shot and killed (it remains unclear by whom). Both the Quebec government and the Mohawks established more blockades to try to pressure the other side: the SQ trying to isolate the Mohawk resistance and the Mohawks trying to escalate public pressure by blockading the Mercier bridge, which connects Montreal to its southern suburbs. The federal government was deeply involved: the Royal Canadian Mounted Police were brought in, and 4,000 members of the Canadian Forces were deployed. As the

Figure 7.1. Oka Standoff
Source: The Canadian Press/Shaney Komulainen.

summer wore on, solidarity protests and blockades were mounted by Indigenous Peoples across Canada, and as many as 10,000 people taking to the streets of the Montreal suburb of Chateauguay to protest the blockade of the Mercier bridge made national headlines.

The outcome of the Oka Crisis or Kanesatake Resistance was mixed. Eventually, the federal government bought the land in question, cancelled the golf course expansion, and bought other land for the Kanesatake Mohawks, although this was still short of the full amount of land they claimed and was land ownership on the state's terms, not a recognition of Mohawk sovereignty. Thirty-four people were arrested when the Mohawks surrendered on September 24. The following year, the federal government launched the Royal Commission on Aboriginal Peoples, which published its landmark report in 1996 (Dussault et al., 1996). At issue in this conflict was not only land as a foundational resource, one that is necessary for access to other resources, but also the question of how land is and should be conceptualized, whether land ought to be understood and organized as a resource in the first place.

7.1 INTRODUCTION

In previous chapters, we have shown how the conceptual transformation of an ecosystem component into an abstract resource facilitated the material extraction of the resource from its physical location, for example, understanding forests as lumber or fossilized carbon as fossil fuel. In the case of land, this happens when it is conceptualized as soil or mineral resources. This abstracting, extractivist mentality is part of what drives the monumental landscape transformations of the Anthropocene (see Box 2.2). The significance of land as a resource, however, is much broader than that. As with water in the previous chapter, land can also be seen as a component of the landscape that facilitates the extraction of other resources. The extraction of forestry or fossil fuel resources depends not only on the conceptualization of trees or coal as a resource but also on seeing as a resource the land where those resources are situated, such that access to timber or fossil

BIG IDEAS IN SMALL BOXES

BOX 7.1. SETTLER COLONIALISM

Settler colonialism refers to a set of practices and of ideas. If colonialism as a general practice refers to territorial expansion or conquest, settler colonialism specifies that a large number of residents from a metropolitan country move to (settle in) the area that is being colonized. Thus, rather than ruling over Indigenous populations, settler colonialism is a form of colonialism in which colonizers seek to displace or replace the Indigenous inhabitants of a territory. Along with Canada, the United States, Australia, and New Zealand (among others) are examples of settler colonial countries.

Colonizers justify their right to settle a new territory in a couple of ways. At times, they assert that the land they seek to occupy was, if not uninhabited, then at least ungoverned according to European understandings of international law. European legal doctrines, such as the doctrine of discovery or the idea of *terra nullius* (literally, land belonging to nobody), were invoked to claim the right to stake sovereign authority over a piece of land. In the *Tsilhqot'in Nation v. British Columbia* (2014) case, the Supreme Court of Canada ruled that the Royal Proclamation of 1763 confirmed that the doctrine of discovery did not apply in Canada. The Truth and Reconciliation Commission of Canada's (2015) Calls to Action include a recommendation (no. 47) that the Government of Canada repudiate the doctrine of discovery.

A distinct but related set of justifications is framed in developmental terms: the colonizers claim to be more advanced than Indigenous inhabitants, and thus colonization is what allows the colony to progress or develop. An influential example of this is found in the arguments of English political theorist John Locke (1632–1704). Referring specifically to America (without having lived there), Locke argued that private property in land, and the inequalities that might be generated, was justified because it allowed the land to be used more productively. According to Locke, America (recalling that British North America at the time included parts of what is now Canada and the United States) was a land of "wild woods and uncultivated waste" (Locke, 1690/1980, Section 37).

It is important to note here that Locke (1690/1980) is referring to the improvement or development of the land rather than people. Again, this distinguishes settler colonialism from other forms of colonialism. Whereas other forms of colonialism are sometimes justified in terms of civilizing or bringing development to colonized peoples, they are at least in principle time limited. Over time, the colonial mission should succeed, at which point the colonized people would be sufficiently developed to govern themselves, or so goes the thinking. Of course, in practice these colonial relations were only ended by struggles (often violent) waged by the colonized. But such an endpoint cannot be envisioned for settler colonialism, because the presence of Indigenous Peoples as Indigenous Peoples is seen as the obstacle to be overcome through colonization. In this sense, Patrick Wolfe (2006) describes settler colonialism as "a structure not an event" (p. 388), meaning that it is an ongoing set of processes rather than something confined to the past (see also Tuck & Wang, 2012).

fuel resources can be controlled. In other words, land-as-resource is necessary for resource thinking more generally. In Canada, the transition from thinking of land as a relational web to thinking of land as owned territory or property, or what Macarena Gomez-Barris (2017) calls the extractive view, has been fundamental to the operation of settler colonialism (Box 7.1).

This extractive view of land is exemplified in the Geological Survey of Canada's mapping of the Canadian landscape in the nineteenth century, which transformed the landscape into settler space in two ways (Martin, 2009). First, these surveys produced concrete information about the landscape – rivers, minerals, soils – that "served to draw the lands, natural resources and Indigenous inhabitants of the West more fully into the administrative orbit of the Dominion government" (Martin, 2009, p. 3). Second,

> these scientific surveys signified Canada's desire and capacity to assert its epistemological dominion over the West. Legal dominion had been established with the land transfers of 1870 and 1871 but owning the territory and knowing it were two different matters. To reinforce Canada's sovereignty over the West it was imperative that the fledgling nation take swift measures to establish epistemological control over its new acquisitions. In this context, as Raymond Craib suggests, the work of a publicly-funded scientific survey was about more than gathering useful information about resources and terrain – *it was also a profound symbol of authority because a state's power to explore and map its national territory signified its power to rule over that territory*. (Martin, 2009, pp. 3–4, emphasis added)

Of course, the idea of human (settler) separation from and mastery and control over nature did not go unchallenged. Many Indigenous traditions and understandings of the world (ontologies) are centred on land. For example, Yellowknives Dene scholar Glen Coulthard (2014) explains how land is understood both as a place and as "a way of knowing, of experiencing and relating to the world and with others" (p. 61). Coulthard continues, "In the Weledeh dialect of Dogrib (which is my community's language), for example, 'land' (or *dè*) is translated in relational terms as that which encompasses not only the land (understood here as material), but also people and animals, rocks and trees, lakes and rivers, and so on. Seen in this light, we are as much a part of the land as any other element" (p. 61).

Similarly, in her writing about Nishnaabeg conceptualizations of land (*aki*), Leanne B. Simpson (2014) writes that "aki includes all aspects of creation: land forms, elements, plants, animals, spirits, sounds, thoughts, feelings, energies and all of the emergent systems, ecologies and networks that connect these elements" (p. 15).

Criticisms of the reduction of land to resource can also be found among ecological thinkers working in Western settler traditions. For example, Aldo Leopold (1966), a twentieth-century ecologist who worked for the US Forest Service, argued that land should be understood as a "community" of which human beings are but one member: "Land, then, is not merely soil; it is a fountain of energy flowing through a circuit of soils, plants, and animals" (p. 253). On the basis of this understanding of land, he argued for the cultivation of a "land ethic," in which human beings would move from "conqueror of the land-community to plain member and citizen of it" (p. 240).

For Leopold (1966), the absence of a land ethic means that human decision makers are focused on their short-term economic interests, ignoring the complexity of the land community and the role played by (and hence the value of) all members of it. Ignoring this complexity threatens ecological (and economic) ruin in the longer term. Similarly, noting the loss of traditional ecological knowledge (her examples include children's inability to identify more than a handful of local plant species), Robin Kimmerer (2014) of the Citizen Potawatomi Nation suggests that humans are being brought "to a condition of isolation and disconnection, that philosophers have called 'species loneliness.' Species loneliness – this deep, unnamed sadness – is the cost of estrangement from the rest of creation, from the loss of relationship. As our dominance has grown, we have become more isolated, more lonely on the planet, and we can no longer call our neighbors by name" (p. 21).

Thus, although the conceptualization of land as a resource facilitates the extraction of other resources and hence, as we have shown, economic growth and nation building more broadly, it also comes with costs. Because resource thinking dominates most current political life in Canada, making up Canadians' "common sense," those costs are rarely acknowledged. Arguably, the most significant costs of settler-driven nation building have been borne by Indigenous Peoples. Resistance to these historical and ongoing injustices have catalyzed movements such as Land Back, whose aim is to "get Indigenous Lands back into Indigenous hands" (Land Back, n.d., para. 1), and Idle No More (n.d.), whose vision states, "We must repair these treaty violations, live the spirit and intent of the treaty relationship, work towards justice in action, and protect Mother Earth." Both movements recognize and seek to address the broken treaties and *terra nullius* narratives that laid the foundation for building Canada.

In this chapter, we examine three different cases of the development of land-as-resource: agriculture and the settlement of the prairies, the development and expansion of national parks, and patterns of urban development. Each of these represents a different way of conceptualizing of land as resource. First, and most like the case of water resources discussed in the preceding chapter, land is imagined as soil of a particular quantity and quality. Second, land is seen as an idealized construction of landscape features as a cultural resource or symbol. Finally, land is transformed into something that can be bought and sold (real estate) by imagining it as abstract coordinates in space. As with previous chapters, in the second half of the chapter we analyse these cases through the lens of the channels through which ecosystems and their components are organized into resources.

7.2 SOIL

One way to understand the transition of land from ecosystem component to resource is through soil and, in particular, how specific features of soil have allowed particular kinds of agricultural development. Here, we draw on the research of Julia Laforge and Stéphane M. McLachlan. Laforge and McLachlan (2018) trace how soil, water, and seeds worked together in the Canadian nation-building project to displace Indigenous histories – material and cultural – from the landscape and replace them with the colonial

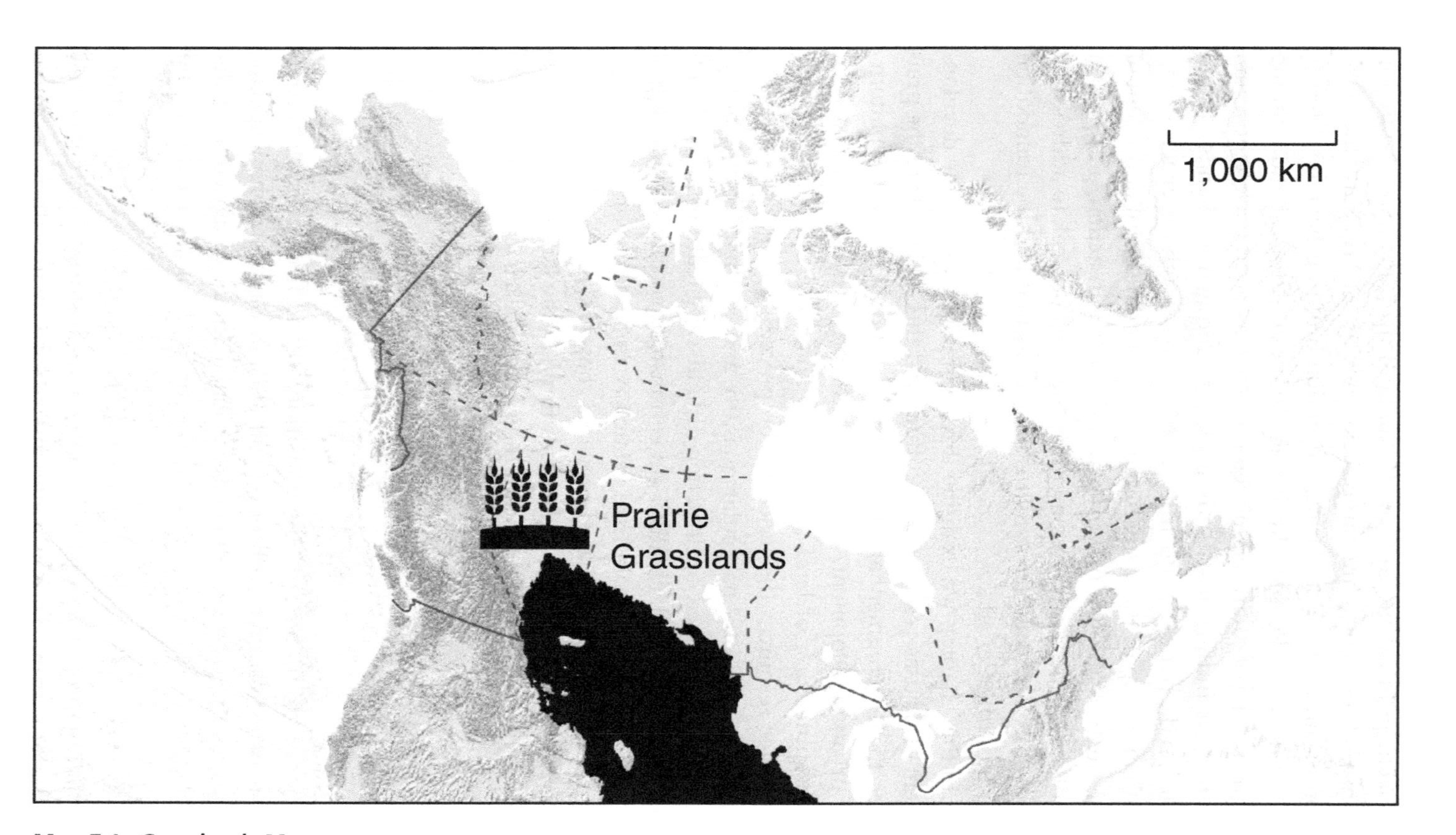

Map 7.1. Grasslands Map

nation-building project of developing valuable crops on the Canadian prairies (Map 7.1). They draw on Foucault's (2009) concept of "governmentality": a kind of rationality (way of thinking) that is specific to governing and is focused on governing (managing, steering) populations (not individuals) with the goal of "improv[ing] the condition of the population" (p. 105). As European settlers and the land they occupied were increasingly seen as objects to be governed, the settlers and their states were transformed over time from new arrivals, to owners, to exclusive managers of the land. This process, in turn, was made possible by the erasure of the presence of Indigenous Peoples and practices, including "the erasure of a culture of agriculture from many Indigenous histories" (p. 363). Laforge and McLachlan (2018) continue, "However, the understanding of the Canadian Prairies as non-agricultural, underutilized land was artificially created as part of the colonial process ... Farming and gardening were practiced by many First Nations and Métis peoples before Euro-Canadian settlement.... However, this agricultural history has been effectively erased from dominant historical narratives, which emphasize an arduous survival based on hunting and gathering" (pp. 363–4).

Through the twin processes of attempted Indigenous erasure and actively encouraging settlement of the (newly) empty space, agriculture became an increasingly important part of the Canadian economy and of colonial Canadian identities. Agriculture has played, and continues to play, a significant role in the Canadian economy: in 2016, the agriculture and agri-food system generated $111.9 billion (about 6.7 per cent) of Canada's GDP and employed 2.3 million people, or 12.5 per cent of the population (Agriculture Canada, 2017). Despite this significant – and growing – contribution, the number of farms in Canada is decreasing while the number of farmed acres continues to increase, indicating that fewer farmers are owning bigger and bigger properties (Statistics Canada, 2017). This shift to fewer and larger farms is a shift from labour-intensive farming to capital-intensive farming: the small family farm where a handful of individuals provide the physical labour to grow and harvest crops is being replaced by larger farms increasingly run through mechanized labour. For example, cows were milked by hand until the introduction of automated milking machines. This process was labour intensive, requiring farmers to spend hours per day hand-milking dairy cows. Because it required little more than a stool and pail (and cow), the capital investment was relatively low. Today, almost all Canadian dairy farms use automatic milking. Automatic milking takes significantly less time than milking by hand, but the equipment is considerably more expensive – hundreds of thousands of dollars. Because this amount is a significant investment, and because investors (i.e., farmers) want to maximize the return on their investment, there is an incentive to purchase more cows so that the equipment is working to full capacity (a single cow can be milked two to three times per day). More cows mean more barn space and feed, and added to the cost of the milkers, the start-up and operating costs of a dairy operation can become prohibitively high for entry-level farmers. The introduction of this technology thus facilitates the transition to a smaller number of increasingly automated, capital-intensive, large-scale farms.

As a result of these kinds of shifts across a variety of crops and regions, agricultural land in Canada has gone from being used in a labour-intensive way to being used in a capital-intensive way. At the same time, land and soil have gone from being ecosystem components to being resources that are managed with fertilizers, measured in land surveys, and so on. By simultaneously mechanizing agriculture and reducing it to its physical components, Canadians are, in everyday ways, organizing nature. At the same time, this way of organizing nature organizes them into particular patterns of settlement, land use, and economics.

The agricultural sector was a significant part of building a burgeoning national economy through the nineteenth and early twentieth centuries. Equally important, however, was the sector's role in nation building. Agriculture was one of the central rationales behind the construction of the Canadian Pacific Railway (Box 7.2). In addition to connecting Canada from coast to coast, the railway enabled European settlement of the prairies, and this settlement then became part of the national narrative. An illustrative example of this is the Heritage Minute titled *Soddie* (see Historica Canada, 1991b) depicting Eastern European settlers in the 1890s working hard to build a soddie (a house made of earth) on the Canadian prairies. The clip makes no reference to the Indigenous Peoples who lived there for thousands of

Figure 7.2. "Dig for Victory"
Source: © Imperial War Museum (Art.IWM PST 16807).

years and thus replicates and amplifies the settler colonial discourse that erases Indigenous presence on the land to create a wilderness for Europeans to discover and settle.

A different kind of example of the link between agriculture and nation building occurred during World War II, when Canadian families were encouraged to grow Victory Gardens – that is, vegetable gardens – to make the products of large-scale agricultural production available for soldiers (Figure 7.2). To that end, agriculture and vegetable gardening were framed as patriotic acts, with gardening encouraged as something a "housoldier" (Mosby, 2014) could do to participate in war efforts (Figure 7.3).

BIG IDEAS IN SMALL BOXES

BOX 7.2. THE CANADIAN PACIFIC RAILWAY

The rail line that runs across the country, the Canadian Pacific Railway (CPR), is an example of how resources have been organized. The Canada of 1867 (the time of Confederation) included the four provinces of Ontario, Quebec, New Brunswick, and Nova Scotia. Manitoba joined in 1870. The next province to join the federation was British Columbia, in 1871, and it did so on the condition that the federal government would build a rail line – a physical connection – between British Columbia and the rest of the country. Construction was fraught with both financial and engineering challenges, but the rail line was completed in 1885. Much has been written about the railway elsewhere, but we raise three main points here about how the CPR was central to the development of colonial Canada.

First, the railway organized resources. It allowed goods produced in the Pacific and Prairie provinces to be easily transported to the East Coast and exported to Europe (the primary export market at the time). It also allowed for the flow of goods from one end of the country to the other. This physical connection densified the web of material, financial, and cultural connections between various regions of the country, contributing significantly to a sense of nation building in the emergent country.

Second, the railway also organized people, beginning from its construction, which was grounded in the *terra nullius* concept: the idea that the rail line was being built in the uninhabited wilderness, when in fact much of the land was actively occupied by Indigenous Peoples. For example, some 5,000 Indigenous people were expelled from southern Saskatchewan in the 1880s to make way for the new rail line (Andrew-Gee, 2020), and the CPR runs through many Indigenous territories; this is why some high-profile Indigenous protests continue to be sited on the rail lines. The CPR also organized people through immigration. Of the 9,000 workers who built the railway, 6,500 were Chinese Canadians who came to Canada for jobs on the rail line. The Chinese workers were paid less than white workers and undertook riskier jobs (e.g., handling the explosives used to build tunnels), and hundreds died from accidents, cold, and malnutrition (Government of British Columbia, n.d.; see also Historica

Canada, n.d.). Once the railroad was built, it continued to be used for immigration in bringing European settlers to the Prairie provinces. Under a variety of federal and provincial immigration programs, Europeans were promised cheap or free tracts of land, much of which had been emptied of its Indigenous occupants, given to CPR for construction, and then promised by CPR to European immigrants as a way to attract passengers (CP, n.d.). European settlement of the West is described in more detail elsewhere in this chapter.

Finally, the CPR serves as a cultural touchstone for colonial Canada. From the songs of Gordon Lightfoot to the iconic CPR hotels across the country to Heritage Minutes and books such as *The Last Spike* (Berton, 1971), a number of (dated) cultural icons lean on the railroad and its construction to harken back to a glorified and romanticized version of this ultimate resource organizer.

Figure 7.3. "Attack on All Fronts"
Source: Rogers (1943).

7.3 SYMBOL

A second way of understanding land as a resource is through the example of national parks, which play a starring role in the settler cultural imaginary. Sweeping mountain vistas, waves crashing on a rocky shore, old-growth forests, and pristine lakes all feature prominently in countless Canadian icons, both big and small. They adorn Canadian coins, websites, and tourism materials. Founded in 1911, Parks Canada's mandate is to "protect and present nationally significant examples of Canada's natural and cultural heritage and foster public understanding, appreciation and enjoyment in ways that ensure their ecological and commemorative integrity for present and future generations" (Parks Canada, 2022, para. 1).

Although Canada's national parks are highly celebrated and visited – hosting over 16 million visitors in 2019–20 (Parks Canada, 2021) – their establishment can be understood as part of a broader narrative about the relationship between people and place and about the ways in which land went from being an ecosystem component to a cultural and economic resource. In this case, the resource in question – a national park – is not something that is extracted like fish or timber. Instead, it is a resource that generates revenue through tourism and, arguably more importantly, serves as a cultural resource by acting as a mental and emotional touchstone of national identity. Given the prevalence of national park imagery across a broad range of Canadian visual media, from television commercials for Canadian brands to postage stamps, it is fair to say that Canada's national parks have been resourcified as iconic Canadian wilderness. Because these images are ubiquitous, Canadians can participate in or consume national parks even if they never set foot in them. More important, Canada's parks are framed as places of recreation, where busy (urban) Canadians can get away from it all and reconnect to their wild roots, and where nature is protected. This framing reinforces the nature–society binary that we critique in this book: urban spaces are not devoid of nature, nor are natural spaces such as parks unaltered by human impacts.

Indeed, "stripping nature of a resident population had been a staple for the development of parks across large sections of the globe" (Rudin, 2011, p. 163), including in Canada. To create these natural

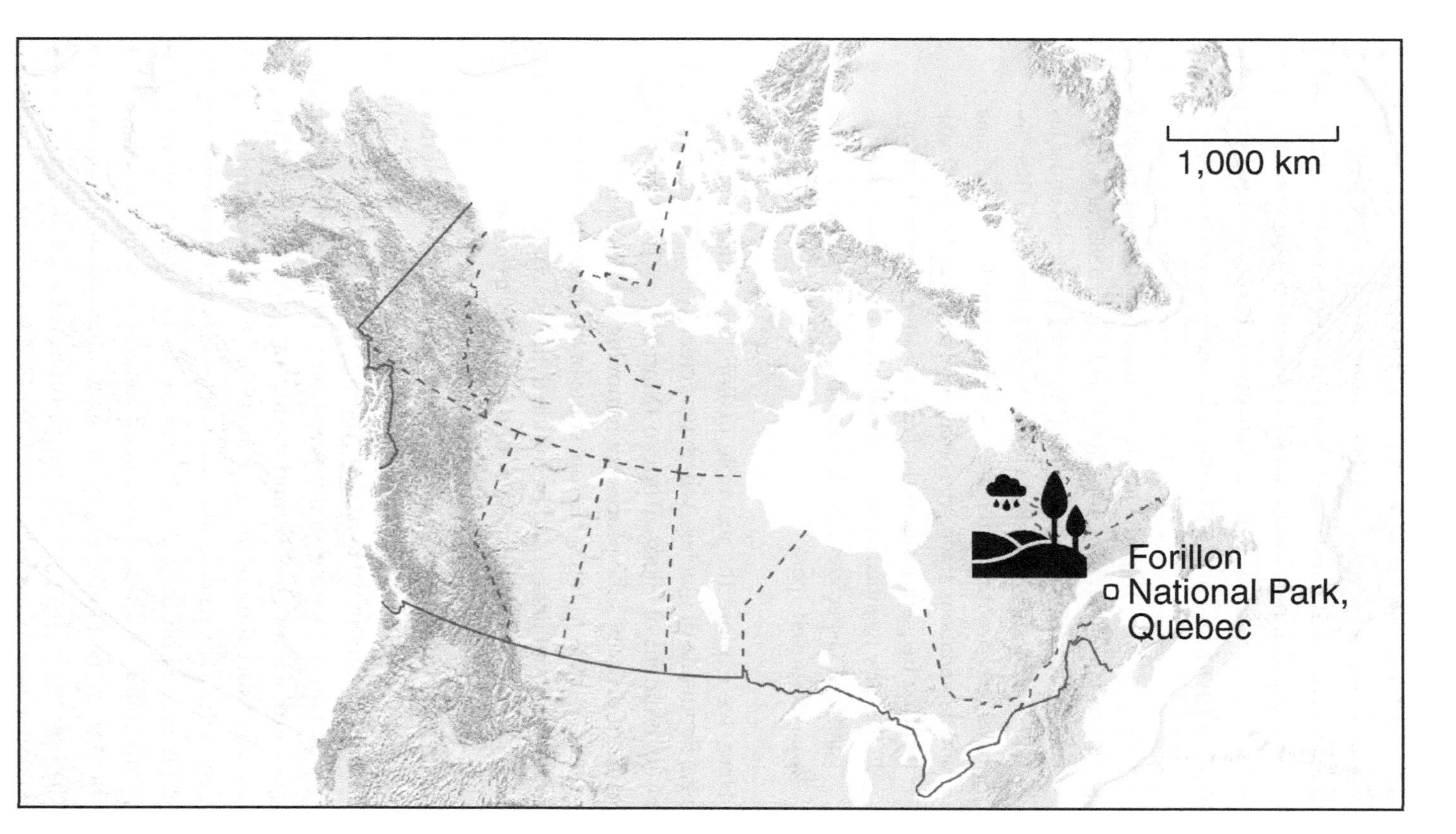

Map 7.2. Parks Map

spaces for Canadians to enjoy, the Canadians living there first had to be removed. Robert Jago writes about this in a 2020 piece titled "Canada's National Parks Are Colonial Crime Scenes," which documents the expulsion of Indigenous Peoples from their own territories in what is now Stanley Park in Vancouver, Banff National Park, Algonquin Provincial Park, and others. In Jago's (2020) words,

> Canada's Parks Departments have treated Indigenous peoples like an infestation ever since the founding, in 1885, of what is now Banff National Park. Looking out at the territories under his supervision, superintendent George Stewart demanded that "Indians" be barred from the area. "Their destruction of the game and depredations among the ornamental trees," he wrote, "make their too frequent visits to the Park a matter of great concern." In the early twentieth century, as non-Native settlements such as Banff and Jasper grew, the livelihoods of First Nations peoples were destroyed. (para. 7)

Here, we explore the example of Forillon National Park, on the Gaspé Peninsula of Quebec, which was created in 1970 under the Liberal government of Pierre Trudeau (Map 7.2). A 1987 Parks history document describes the process of removing the area's inhabitants to create the park as follows: "Before some of the major developments in Forillon National Park could be initiated, the evacuation of park lands within the newly established boundaries had to be arranged. Altogether, about 2,500 separate parcels of land were involved and slightly more than 205 families were affected.... In addition to cash settlements for expropriated land and buildings, former residents were entitled to a relocation grant of up to $2,000" (Lothian, 1987, p. 138).

Rudin (2011) documents the years-long struggle among residents, the Government of Quebec, and the federal government with respect to appropriate compensation for those families whose land had been expropriated. The result of these struggles was that those citizens were forcibly moved off their property to create a public space where Canadians could celebrate their shared natural heritage. Despite a 2011 resolution in the House of Commons to offer an "official apology to the people whose properties were expropriated to create Forillon Park for the unconscionable manner in which they were treated" (Rudin,

2011, p. 161) and a change in Parks Canada policy regarding the way in which parks are established, the vast majority of Canadian parks were created using similar forms of structural violence.

The history of Canada's national parks follows the trajectory of Canadian environmentalism, exemplifying and (more recently) challenging the nature–society binary. What started as an effort to protect iconic areas from development and allow Canadians to "enjoy the benefits of outdoor life and healthful recreation in surroundings of great natural beauty" (Lothian, 1987, p. 29) has become a network of 39 parks covering some 320,000 square kilometres. Particularly in its earlier phases, the transformation of inhabited spaces into "natural" ones involved the removal of permanent human occupants. This recalls what James Scott (1998) calls the ideology of high modernism, discussed in Chapter 4. In the case of national parks, Rudin (2011) notes, the application of high modernism meant that the state had the exclusive right "to speak with the authority of scientific knowledge about the improvement of the human condition and to disallow other competing sources of judgment.... At its most radical, high modernism imagined wiping the slate utterly clean and beginning from zero" (Scott, 1998, as quoted in Rudin, 2011, p. 165). The story of Forillon National Park exemplifies this mentality of needing to clear humans out of nature in the public interest of preserving it. More recently this exclusionist mentality has shifted: some of Canada's more recent parks are co-managed with Indigenous communities, and Canada's newest parks are in areas with very few permanent residents or, in the case of Rouge National Urban Park in Toronto, embedded in an urban area (https://parks.canada.ca/pn-np/on/rouge). Moreover, Parks Canada has a focus on accessibility and public education, reflecting a mandate change from one that was exclusively preservationist to one with increased emphasis on encouraging and facilitating Canadians' engagement with "their" parks, including outfitting parks with amenities such as WiFi, accessible bathrooms, educational programming, interpretation centres, and cultural events.

Once simply part of the landscape, some parts of Canadian land have been transformed into parks. These parks serve as symbols of national identity, of exemplary Canadian nature (however defined). To this end, Canadians have not only organized resources but also, through their organization, created them, and these new symbolic resources have also organized Canadians. Parks Canada has 4,000 employees managing and promoting the country's parks; towns in and near to the parks rely heavily on tourism for their economic sustainability; Canadians plan trips in and around the parks and spend their financial resources on these visits. As such, the parks symbol has had very concrete implications in terms of the mutual development of people and place.

7.4 SPACE

A third example of land resourcification is the case of real estate: areas with clearly defined boundaries that can be bought, sold, and rented. As noted earlier, fundamental to the treatment of land as a resource is the capacity to think of it in abstract terms, as points in space. Key to this way of thinking is the (high modernist) belief that land can be made "legible ... from above and from the center" (Scott, 1998, p. 2). The idea that land can be made legible implies two things: first, that it is possible for the state (or whatever agency is at the centre; in Canada today large corporations in the financial and technology sectors, among others, might also be included) to know all it needs to about a distant community. Second, it also implies that the central agency plays an active role in reshaping the land to make it legible.

Scott (1998) describes the cadastral map – a map that provides "a more or less complete and accurate survey of all landholdings" – as "the crowning artifact" (p. 36) of the project of turning land, understood as a complex and dynamic set of socioecological relations that define a particular place, into an abstract resource (real estate). In European settings, this project was imposed on already existing communities. In Canada, it was imposed on what was described as *terra nullius*. So, although modernizing European states understood that they were trying to create a simple and legible order out of a diverse array of complex and often resistant communities, the state in Canada believed itself to have a blank slate where it could create a simple and legible order from the beginning. In Ontario, for example (and unlike the United States in this regard), land was surveyed and parcelled before it was allocated to settlers. In some areas, roads that provided settlers access to newly surveyed land were (and still are) called "concession roads" or "colonization roads." Because the Crown's right to dispose of (concede to settlers and indeed to colonize in the first place) these lands was disputed, the names have a particularly poignant resonance, a point clearly made in Michelle St John's (2016) documentary *Colonization Road*.

The largest example of this is the Dominion Land Survey, which, starting in the early 1870s, just a few years after Confederation, surveyed and imposed a grid on land from Winnipeg to the Rocky Mountains and from the US border to the North Saskatchewan River (see Map 7.3; note the division of much of the lower part of the map into evenly sized squares). The pattern remains visible from the air in the checkerboard pattern of the prairie landscape. As an early part of establishing the structure of settler colonialism, the survey was predicated on Indigenous erasure and dispossession: the presumption was that Indigenous Peoples did not (or at least did not rightfully) inhabit the landscape. As such, once surveyed, it could be populated with settlers who were induced by the government to move there with the promise of land ownership, work opportunities, and monetary payments (Figure 7.4).

Another example of this imposition of order onto assumed blank space is the case of Toronto and its downtown street grid, which dates back to the late eighteenth century (Map 7.4). The structure is replicated beyond the downtown core because most major arterial roads follow a similar grid pattern. Although eighteenth-century planners certainly did not envision a city with millions of car-driving inhabitants, rural properties around the early city were nevertheless built around regularly spaced concession roads. Thus, major east–west roads in Toronto – Bloor, St. Clair, Eglinton, Lawrence, and so forth – run parallel in 2-kilometre (1.25 miles as originally measured in the imperial system) increments.

The pattern of settlement and landscape transformation in Canada was typically as follows: The Crown assumed sovereignty over a piece of territory, mapped and surveyed it, built infrastructure (road and rail) to facilitate movement through it, and then distributed parcels of that territory to settlers. As the populations of nearby cities increased and automobility facilitated individual movement over greater distances, rural landscapes were increasingly transformed into suburban or exurban landscapes: sprawling,

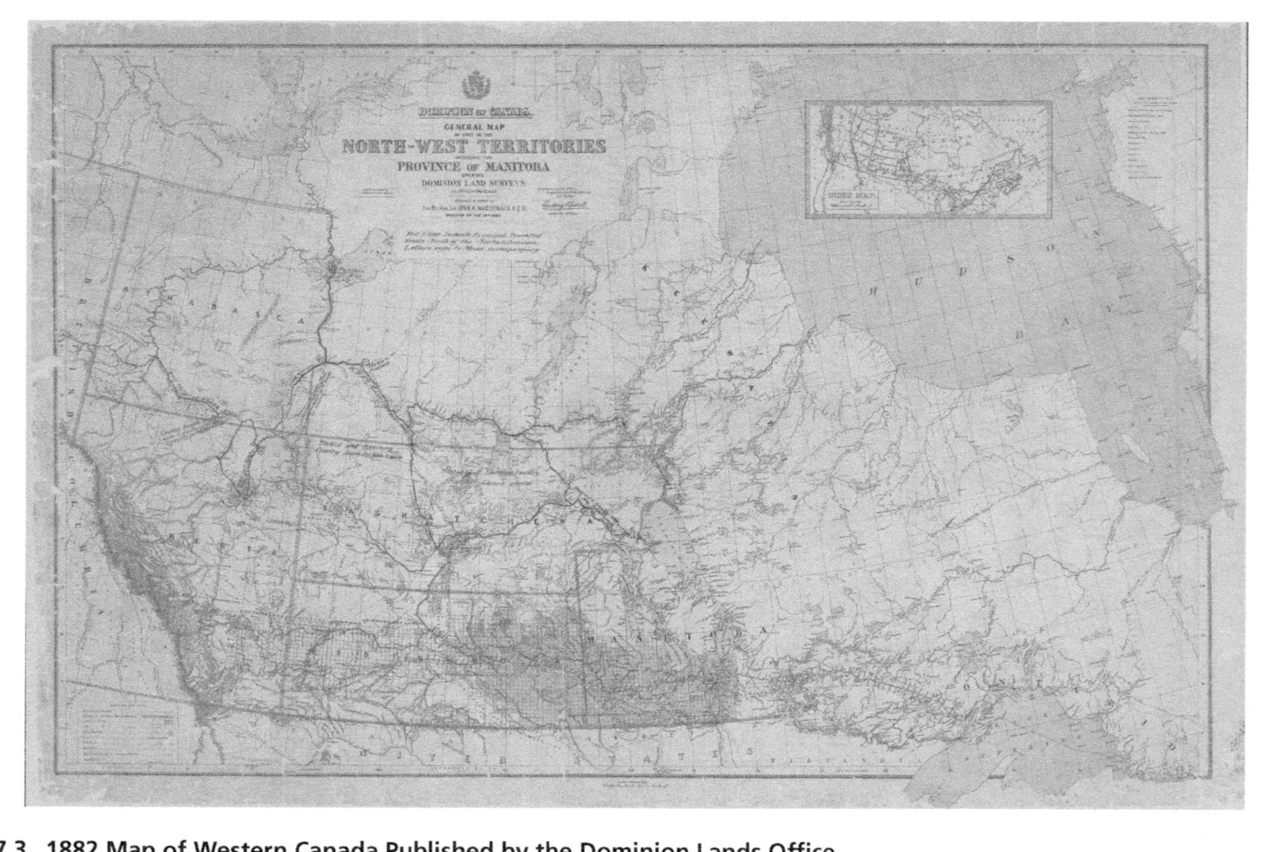

Map 7.3. 1882 Map of Western Canada Published by the Dominion Lands Office
Source: Johnston (1882). Library and Archives Canada/e011309128 (Local Class No. H1/701/1883, Box No. 2000229203).

Figure 7.4. Canada, the Nearest British Colony
Source: Library and Archives Canada/OCLC 1007445917/e010693826. https://data2.collectionscanada.ca/e/e428/e010693826-v8.jpg

low-density residential settlements punctuated with car-centric commercial and office developments. This pattern of land use is of course not unique to Canada, and it increasingly defines the human-built environment globally (Keil, 2017; for an early analysis of the political implications of suburbanization in Toronto, see Dale, 1999). Even more recently, as information is increasingly treated as a valuable commodity in its own right, resource thinking turns places (again, defined by complex and dynamic socioecological relations) into a repositories of location-specific data. Sidewalk Toronto – a public–private partnership between the city and Alphabet Labs (owned by the same parent company as Google) – proposed the construction of "an urban district that is built around information technology and uses data ... to guide its operation"

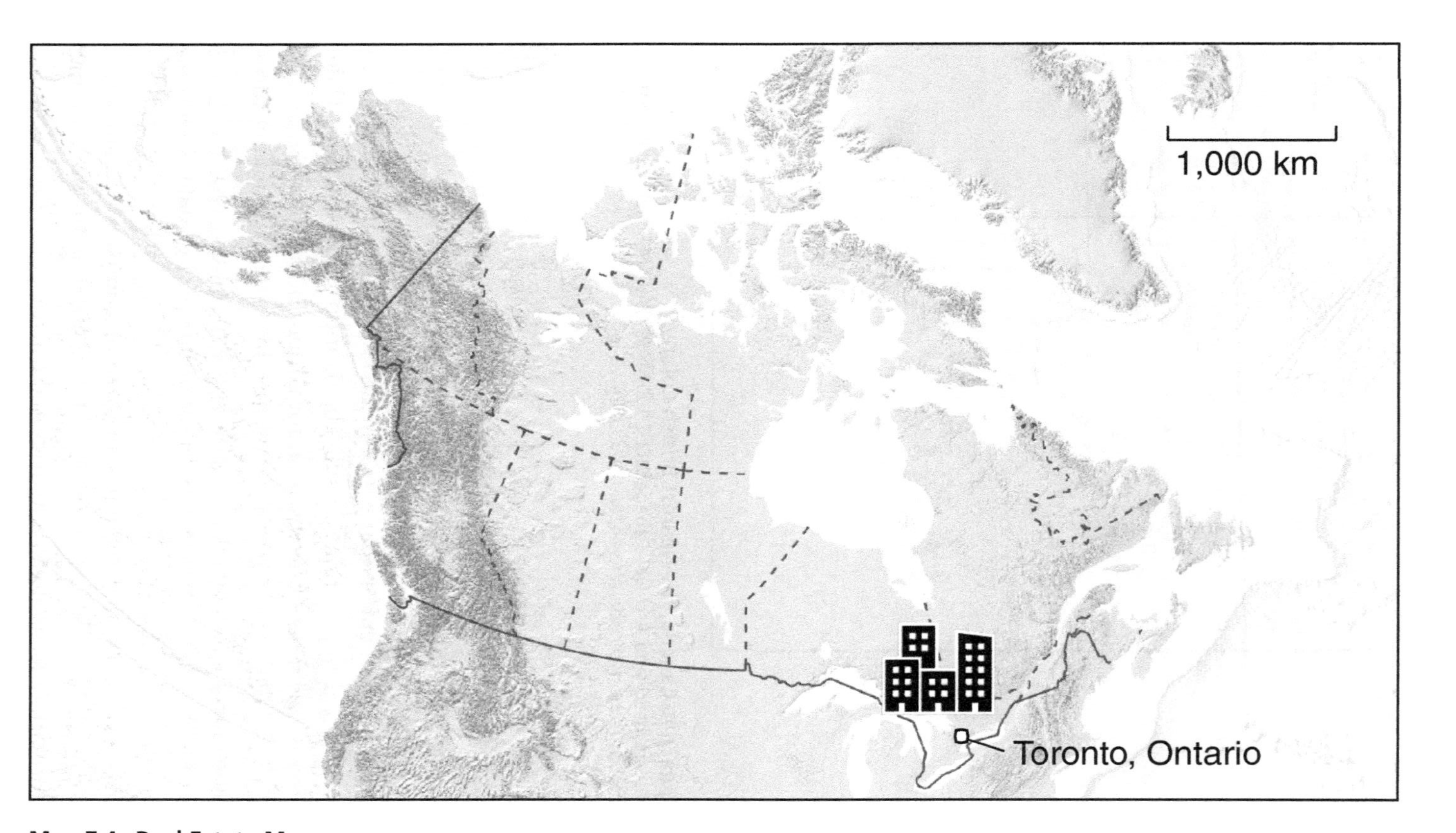

Map 7.4. Real Estate Map

(Bozikovic, 2017, para. 4). Its "smart city" vision included the installation of "sensors and other data-collecting devices to monitor things including pedestrian traffic, noise, weather conditions, and energy and garbage use" (Vincent, 2020b, para. 3). That project was cancelled in 2020, according to its proponent, because of the economic effects of the COVID-19 pandemic (Vincent, 2020a). However, up until the point of its cancellation, the project remained controversial because of concerns about the collection, storage, and governance of massive amounts of citizens' data. Although proponents frame such projects in terms of the promise of optimization (traffic will run more fluidly, energy will be used more efficiently, residents and business will generate less waste, etc.), critics note that Google and other similar massive information technology companies (Facebook, Microsoft, Apple, Amazon) have a proprietary interest in using such data to optimize their ability to deliver targeted advertising (Zuboff, 2019).

Both the history of suburban development and the redevelopment of urban cores (as in the Sidewalk Toronto case) exemplify the power of real estate developers to (re)shape landscapes. Although in some ways the powers of municipal governments are quite limited compared with those of federal or provincial governments, they do have the power to regulate land use in the places where most Canadians live and work. As Robert MacDermid's (2009, pp. 26–33) study of Toronto indicates, real estate developers are, by a large margin, the largest contributors to municipal election campaigns. At the same time, municipal governments have at their disposal very limited tools to generate revenue, and by far the most significant of these is property taxation. So municipal governments have a strong interest in increasing property values (higher property values means higher property tax revenues, even if the tax rate does not change), and they can do so through zoning changes that produce benefits for real estate developers. "Most of the profit for land developers is captured in the rezoning, subdividing and servicing of raw land.... Almost all of a developer's [*sic*] profit comes from the planning phase, where municipal politics creates wealth through land use planning, servicing and subdivision into smaller lots" (MacDermid, 2009, pp. 33–4). Here, too, Canada mirrors a global trend whereby urban land is increasingly seen through the lens of its potential to be a profitable investment:

"housing is becoming ever less an infrastructure for living and ever more an instrument for financial accumulation" (Madden & Marcuse, 2016, p. 18).

In the preceding four chapters, we discussed resources that could be picked up and moved. In all four cases, at least part of the story we told was one in which the resource was extracted in an increasingly intensive way over time. Because land cannot be picked up and moved, the story in this chapter is somewhat different, although there are parallels. Land as soil hews closest to the previous chapters, because soil does move (via erosion), and soil quality can be degraded from overuse. Although parks can be overused in the sense that too many human visitors may risk spoiling the natural or wilderness quality of the landscape, such problems do not exist for the circulation and consumption of park images. Instead, the reliance on mediated rather than immediate experience (looking at images rather than visiting the park itself) allows for a disconnection: experiencing or consuming the land without being on the land. This disconnection, which resonates with Scott's (1998) idea that high modernism involves seeing "from above," is also found in the conceptualization of land as real estate. The increasing treatment of land as speculative real estate investment shows how, even if it cannot be extracted more intensively, it can be abstracted more intensively.

These examples show how land, broadly defined, went from being an ecosystem component that was part of the Indigenous landscape to being a resource, as soil, symbol, or space. That transformation was enabled through the various channels that organize resources, which we turn to next.

7.5 CHANNELS IN ACTION: ORGANIZING LAND

We begin by considering state institutions as a means by which resources are organized and those resources are used to organize people. Perhaps even more so than in previous chapters, however, our discussion of the **governments** channel also shows how these channels are interwoven and dependent on one another. Indeed, in the case of land in particular, one can see how a specific understanding of land-as-resource that dominates over other understandings of land in the realm of **culture and ideas** undergirds the changes that flow through the other channels.

Although land-as-ecosystem-component generates innumerable ecosystem services (see Chapter 4), managed land – that is, land-as-resource – generates capital, which is a central interest of the colonial state. Land-as-resource generates capital by facilitating extraction or development. That is, it is not really the land itself that is economically valuable, but rather what can be done on the land. Drawing on our earlier examples, land can be used to grow crops or raise dairy cows, it can be a tourist attraction, or it can be the foundation on which real estate developments are built; it can also be used to grow trees for forestry activity or as terrain onto which roads, airports, and railways can be built. In short, it is not the land itself that is of interest, but what the land can be used for.

Because land can be used for an almost infinite array of revenue-generating activities, the state has an interest in its control. Indeed, one of the first priorities of colonization was the installation of a European system of property ownership – one in which individuals and governments can own tracts of land and have exclusive control over their development, use, and access. This ownership model (Singer, 2000, as cited in Egan, 2013), in which plots of land are clearly delineated and can be sold or traded by the owner, is antithetical to many Indigenous relationships to land (see, e.g., L.B. Simpson, 2017). As many historians and scholars have noted, "Colonialism relies on the displacement and dispossession of Aboriginal peoples to open up lands for European settlers" (Egan, 2013, p. 37). This displacement and dispossession was accomplished through, first, a world view and narrative that (incorrectly) saw Indigenous territory as wild, savage, and chaotic, with no system of ownership or organization, and second, the colonial imposition of "order" on "chaos" through cession treaties in which the Crown became the landowner (Miller, 2009). State ownership of land was central to Canada's colonial nation-building project, which was predicated on (1) the notion of a *terra nullius* welcoming to European immigrants and (2) facilitation of the types of resource development on which an emergent Canadian economy was dependent: agriculture, forestry, and later mining. This settler colonial project has been resisted by Indigenous Peoples from the beginning, including most recently in the Idle No More and Land Back movements.

Thus, since Canada's inception as a country, the state and land have mutually organized each other. Because some areas are drier than others, some areas are forested and some are less so, some areas are good for agriculture and others less so, and so on, Canadians have grouped themselves – and their institutions – accordingly. Similarly, Canadians – and Canadian institutions in particular – have organized land in a variety of ways. As a property owner, the Canadian state was quick to use its institutions to set up systems for governing land, particularly with respect to the examples discussed in this chapter. With respect to soil, Agriculture Canada was founded a year after Confederation, in 1868, and, as discussed in Chapter 6 of this volume, control of water was (and is) central to the development of land as a resource: for example, irrigation districts in Alberta's agricultural south were first authorized in 1894 (R.F. Smith, 1978). Land as symbol – that is, parks – is mediated through Parks Canada, which is a division of ECCC. Land as space – that is, real estate – is managed through a number of institutions, among them the Canada Mortgage and Housing Corporation (CMHC), which was established as a Crown corporation in 1946 to help soldiers returning from World War II access quality housing. CMHC has played a critical role in Canada's real estate market since its inception. More recently, federal government policies that have made it easier for immigrants to obtain mortgages have also had the effect of "creating highly-leveraged [indebted] immigrant and visible minority households and neighbourhoods" (Simone & Walks, 2019, p. 296). The complex and contradictory character of official multiculturalism is a significant part of Canadian identity, and one that both reproduces and sits uneasily with inequality in its various dimensions. Many other examples could be cited: all municipalities have zoning bylaws controlling which spaces within a community can be used for residential, commercial, or industrial purposes; the *Indian Act* governs land use on Indigenous reservations; Canadian Heritage controls the way in which historical buildings can be managed; Transport Canada is responsible for highways, rail lines, and airports that are both ubiquitous throughout and provide connections between points on the Canadian landscape; provincial departments approve (or deny approval of) mining and forestry operations that can significantly alter landscapes; National Defence manages lands on military bases across the country; and the list goes on. The point here is that federal, provincial, and municipal governments are invested in the control of land, in large part because the management of land and its conversion from ecosystem component to resource is central in the development of (colonial) national identities, communities, and economies. At the same time, Indigenous Peoples have their own systems of governance and relations with the land, which often conflict with settler institutions (Pasternak, 2017).

Let's turn to look at communities as channels for organizing land, and vice versa. Canada's **communities** have been central to the processes through which people and nature have organized one another, through both settlement patterns and the establishment of resource communities.

As discussed previously, the processes of settler colonialism established communities in strategic areas. In some cases, this involved the displacement of Inuit Peoples into remote areas to act as "human flagpoles" as a symbol of Canadian sovereignty in the north (Arnold, 2008). This is another example of nation building that was predicated on occupying the *terra nullius*. In other cases, communities were established in resource-rich areas so that the people living in those communities could actively participate in – and gain economically from – the conversion of landscape into resource.

We have described several instances in previous chapters of communities being built around resources on or under the land: the forest resources in and around Miramichi, New Brunswick (Chapter 4); and coal in Sydney, Nova Scotia; and mines and bitumen in Fort McMurray, Alberta (Chapter 5), for example. To this we could add other resources, such as minerals, that are not the focus of specific chapters in this book. There is Dawson City, Yukon Territory, site of the 1896–7 Klondike gold rush that turned the fishing site of the Tr'ondëk Hwëch'in First Nation at the confluence of the Klondike and Yukon rivers into a town bursting with some 40,000 miners determined to strike it rich. There is Sudbury, Ontario, an active mining community and home to the iconic Big Nickel (see Figure 7.5). There is Asbestos, Quebec, named for the fibrous mineral that formed the backbone of the town's economy until markets for the substance dried up in the wake of overwhelming medical evidence of its toxicity.

In all these cases, the extraction of the specific resource in question depended not only on the conceptualization of that thing (timber,

Figure 7.5. The Big Nickel
Source: Photograph by Andrew Biro.

bitumen, nickel, etc.) as a resource but also on the conceptualization of the land itself as a resource. The treatment of land as property – accessible to some and not to others – is an essential precondition for all these extractive projects. These extractive projects in turn necessitate the development of communities to support them. They bring together the people who do the extractive work, as well as those who support them in one form or another. Fort McMurray is an oil community, but it is also a small city with 60,000 people (not all oil sands workers), schools, a college, restaurants, a hospital, community centres, museums, and so on.

Thus, in addition to physical settlement patterns, land and its transformation has shaped how **economies** (who does what, who gets what, and what they do with it) are organized in particular places. For example, people are paid to work as coal miners only in places where there are significant coal deposits and as farmers only where soil and climatic conditions make agriculture possible. Landscape features not only shape a material economy of goods such as coal or wheat, they also shape a symbolic economy, one that constructs an idea of nature as a resource not for extraction but for discursive reproduction, and one that cashes out values on its own logic. Both through the circulation of iconic images and through Parks Canada's emphasis on engaging Canadians, national parks can be thought of not only as places where nature is preserved, but also as places where it is consumed and memories are produced (A. Wilson, 1991). Rather than parks as places apart from the human economy, parks are a part of the human economy. Thus, park tourism can be seen as a memory-generating industry that employs 4,000 people (and many more non-humans!) to serve 16 million customers per year.

The preceding examples should not be interpreted as an either–or opposition between the organization of material things on the one hand and ideas or values on the other. All economies have both material and symbolic dimensions. Memory generation is a material process, even if it happens largely inside an individual's body, and landscapes within and outside of parks are transformed to facilitate people's experience of nature. Conversely, physically transformative activities such as farming and coal mining produce distinctive cultural values and attachments (the latter memorialized, e.g., in the Cape Breton Miners Museum in Glace Bay, Nova Scotia; https://www.minersmuseum.com/). There is work and struggle involved both in physically transforming the landscape and in (re)shaping how one thinks about and (de)values the landscape.

As we have shown in previous chapters, particular forms of economic organization are enabled by and in turn empower particular actors. The ownership model for land described here has been particularly important for the establishment of settler colonialism and its extractive economies. This drives changes to the **built environment**: patterns of development, including expropriation of the lands inhabited by Indigenous communities (Coulthard, 2014, Chapter 2) in outlying areas or the hinterland (literally, "the land behind"). Perhaps less visibly, it also provides a logic for decisions about how urban landscapes are transformed over time. Over the past few decades, the continued dominance of the ownership model has solidified the view that urban space is and should be composed of commodified real estate. This can be contrasted with the "right to the city" view that "the freedom to make and remake our cities and ourselves" (Harvey, 2008, p. 23) should be allocated democratically rather than on the basis of property ownership (Lefebvre, 1996). Writing about global trends, urban studies scholars David Madden and Peter Marcuse (2016) conclude that "in many places real estate has become more profitable and important than industry" (p. 18). This observation applies to the speculative property booms that have driven growth in Canada's urban centres. Especially (but not only) in Toronto and Vancouver, real estate values have increased spectacularly over the past few decades. Although this makes homeowners wealthier on paper, it is difficult for many to realize this wealth if this real estate investment also serves the essential function of providing necessary shelter for the owner. More than 80 per cent of the worth of property that is not a principal residence is owned by the richest 20 per cent of Canadian households (Rozworski, 2018). Thus, the ownership model of land serves to reinforce economic inequalities among Canadians, first, between those who own property (homeowners) and those who do not (renters and homeless people) and, second, between those whose property consists of their primary residence and those with multiple homes or real estate investments.

The ownership model institutionalizes and empowers certain kinds of economic actors. Rather than just being composed of individuals, the economy is also populated by collective actors. Corporations have many of the same rights as individuals to own and dispose of property, and corporate structures serve as a legal instrument to shield their human owners from liability. Corporations with extensive land access rights (particularly in the mining and oil and gas sectors) and those with valuable urban real estate (including the major banks) are among the most powerful private actors in (re)shaping the Canadian economy.

Although the institution of private property in land ownership can entrench economic and political inequality, public property is not necessarily egalitarian or conflict free. Under the principle of assumed Crown sovereignty, land that has not been designated as private property is nevertheless reserved by the state (the Crown). This encompasses close to 90 per cent of all land in Canada, roughly evenly split between the federal government on the one hand and provincial and territorial governments on the other. As we have indicated in this and previous chapters, this assumption of Crown sovereignty in many cases has been and continues to be contested by Indigenous Peoples, particularly through the Land Back movement, which asserts a right and responsibility to act as a steward of the land in a way that is ontologically at odds with the state's ownership model.

Let us turn finally to **identities**. The Molson commercial that we discussed at the beginning of this book asserts that "it's the land that makes us." The context of the commercial strongly implies that "the land" here means wide-open spaces that bear few obvious human impacts ("more square feet of awesomeness," in the commercial's words). Canadian identities have indeed been shaped by this idealized vision of the landscape, but they have also been shaped by the actual (non-wilderness) landscapes that most Canadians inhabit most of the time.

Writing about the evolution of the Canadian National Parks system, Catriona Mortimer-Sandilands (2009) observes that "particularly following World War I, the presence of rugged, northern wilderness came increasingly to stand in for the national difference between Canada and its 'civilized' British parent. As nature preservation came into prominence in the early twentieth century, the state was charged with the task of developing parks as spaces in which the essence of the Canadian nation could be protected and experienced" (p. 167).

Like the frontier myth described by Cronon (1995) in the context of the United States, there is a long-standing tradition of seeing "authentic" Canadian identity as emerging out of the encounter with powerful natural forces and sublime untouched landscapes. The Molson commercial, and similar contemporary representations, draws on a lineage that stretches back to the beginnings of distinctive settler–Canadian culture. This goes back at least 100 years, to the early twentieth-century landscape paintings of the Group of Seven in visual art: "The 'Canada' that is celebrated as the window into identity is the one painted by the Group of Seven or [one represented in another popular beer commercial], populated neither by native North Americans nor by anyone else who would try to complicate the narratives of identity and territory in Canada" (Manning, 2000, para. 28).

And, again as Cronon (1995) observes in the United States, the idea that national identity is connected to an encounter with rugged, northern wilderness (seen as separate from human society, as the nature–society binary dictates) means that being authentically Canadian is more easily accessible to people of a particular gender, race, or class. Catriona Sandilands (1998) analyses a New Brunswick tourism ad that depicts the province as providing "a pristine wilderness ... for our voyeuristic pleasure" (p. 240). Who are the "we" that are afforded this pleasure? "To take another look at our white, middle-class, heterosexual, Tilley-outfitted couple [shown in the ad], we see that the desire represented in the ad is that of the urbanite. Specifically the fantasy represented/constructed centres on the supposed alienation of contemporary urban life" (Sandilands, 1998, p. 241).

The idealized vision of wilderness landscapes that are presented as essential to Canadian identities are a response to (and escape from) the reshaped (urban) landscapes where most Canadians spend most of their time. These reshaped landscapes, too, affect Canadians differently on the basis of gender, class, race, or other characteristics. Urban environments have been made auto-centric, so cultural amenities or public services are more difficult to access for those who are not able to drive (youth, older adults, people living in poverty). They have been made to include spaces where women do not feel safe walking alone at night or where people of colour are disproportionately subject to police harassment. Rural environments, particularly rural sites of resource extraction and landscape transformation, similarly afford advantages and distribute risks unequally (Gibson et al., 2017).

7.6 SUMMARY AND CONCLUSIONS

In this chapter we have emphasized three interrelated levels of thinking about and treating land as a resource. The first follows most clearly in the footsteps of the preceding chapters. Previous chapters described a shift from treating fish (Chapter 3), trees (Chapter 4), or carbon (Chapter 5) as ecosystem components to treating them as resources that can be conceptually abstracted and physically extracted (fisheries, timber, or fossil fuels, respectively). A similar shift happens in the conceptualization of land as a repository of value, in soil or mineral composition, for example. Here, resource thinking sees land through a lens that renders it quantifiable along various dimensions and thus as essentially interchangeable.

The second level sees land as a symbolic instead of a strictly material resource. Rather than focusing on land as a resource from which value must literally be extracted, here value inheres in the cultural construction and idealization of landscape. The resource is not so much the physical land itself but the ideas and cultural associations that are attached it: natural beauty or links to particular regions or ways of life, for example.

Third and finally, land-as-resource provides a foundation for resource thinking more generally. The idea that land can be abstracted and separated into discrete parcels with clearly delineated borders, as opposed to complexly interconnected and nested ecosystems, is the basis for the commodification of land as real estate. However, it is also the basis for settler colonialism and for the extractive systems discussed in the previous chapters. The forestry industry, coal mining, the extraction of oil and gas, or groundwater, or even the modern fishery system with its territorialized fishing zones, all depend on a system that understands the land on which those resource-extractive activities take place as an inert object that can be understood, controlled, and divided. Of all the cases examined so far, land is perhaps the most abstract and also the most diffuse. Conceptions of land as resource suffuse people's ways of thinking to such an extent that *property* is often used as shorthand for "property in the form of land" (e.g., being on private [or public] property).

In this sense, land clearly illustrates the theme of Indigenous dispossession in its depth and complexity. Settler colonization deprived Indigenous Peoples of the lands that they had occupied for countless generations. Settler colonization is also an ongoing process: Patrick Wolfe (2006) suggests thinking of it as a structure, not an event. Part of that process involves taking a more holistic and relational concept of land and replacing it with one that sees land as abstract and exchangeable property. In other words, it promotes the resourcification of land.

Land as resource grounds Canada as a country. The Canadian state has been instrumental in defining territorial borders, claiming for itself rightful ownership of Crown land and then controlling access to those lands through regulation and systems of private property rights. At the same time, cultural constructions of the landscape have been powerful forces in creating and shaping colonial Canadian national identity. It is difficult, if not impossible, to think of Canada without a prior conceptualization of land as resource.

DISCUSSION QUESTIONS

1. When someone asks you, "Where are you from?" what do you say? Is your answer the name of a town or a city? A treaty area? A watershed? Why do you answer the way you do, and how does your answer reflect the way you relate to the place you are from?
2. Central to the resourcification of land is the concept that pieces of the planet can be clearly defined, sold, and owned. What kinds of ways of life are made possible by this commodification of land, and what ways of life are made impossible by it?

PEDAGOGICAL RESOURCES

Further Viewing

David Suzuki Foundation. (2012). *What is Land Back?* https://davidsuzuki.org/what-you-can-do/what-is-land-back/ [Three-part video series on the past, present, and future of land governance in Canada].

Historica Canada. (n.d.). *Heritage Minutes: Nitro* [Video]. https://www.historicacanada.ca/content/heritage-minutes/nitro

Historica Canada. (1991). *Heritage Minutes: Soddie* [Video]. https://www.historicacanada.ca/content/heritage-minutes/soddie

Obamsawin, A. (Director). (1993). *Kanehsatake: 270 years of resistance* [Film]. Association coopérative de productions audiovisuelles [Award-winning documentary about the Oka crisis].

St. John, M. (Director). (2016). *Colonization Road* [Film]. Decolonization Roads Production Inc.

Further Reading

Canadian Centre for Policy Alternatives. (2019, January/February). *The Monitor*, *25*(5). https://www.policyalternatives.ca/publications/monitor/monitor-januaryfebuary-2019 [Collection of articles that apply the right to the city in Canada].

Indigenous Foundations. (n.d.). *Land & rights*. https://indigenousfoundations.arts.ubc.ca/land__rights/ [Overview of the evolution of Indigenous land rights in the Canadian context, focused on British Columbia].

Miller, J.R. (2009). *Compact, contract, covenant: Aboriginal treaty-making in Canada*. University of Toronto Press.

Neighbourhood Change Research Partnership. (n.d.). Neighbourhood change and building inclusive communities from within. http://neighbourhoodchange.ca/

Yellowhead Institute. (2019). *Land Back: A Yellowhead Institute red paper*. https://redpaper.yellowheadinstitute.org/wp-content/uploads/2019/10/red-paper-report-final.pdf [Report assessing land dispossession and examples of Indigenous reclamation across Canada].

CHAPTER EIGHT

From Bodies to Life

"Paid to carry a stranger's baby – then forced to raise it." That's the headline of a BBC news article (Chong & Whewell, 2018) about the raid of a villa in the Cambodian capital, Phnom Penh, housing 33 pregnant women. The women were carrying the embryos or fetuses of couples from around the world: China, Russia, Canada, Europe, and the United States. For a fee of about US$10,000, the women were to carry these pregnancies to term and deliver the infants to their intended parents (IPs), who had entrusted the surrogates to gestate their child. The plan was a win–win situation: couples who were biologically unable to have a child on their own had a way to do so, and poor women were to be paid the equivalent of a year's salary for the use of their uterus. Instead, in the wake of a new Cambodian law banning surrogacy, authorities raided the facility, arrested the women, and told them that they were responsible for raising the child or they faced 20 years in jail for child trafficking. The outcome was a lose–lose: the impoverished women were on the hook for raising children they could scarcely afford and never intended to meet (often with the stigma of having a child that looked nothing like them or the father), without the US$10,000 they had been promised, and the IPs never met their biological child.

For the IPs, these women (or more specifically, their uteruses) had been resources; for the women, the couples' embryos had been resources. Although the fish and trees discussed in Chapters 3 and 4 are alive, human life does (or seems like it should) belong in a different category, one that is not often thought of as a resource. And yet it is not so different. For some animals, their "liveness" is what makes them a valuable resource: consider hunters or fishers, who are interested in the process of hunting as much as the outcome. Ask a sport fisher to replace a fishing trip with an equivalent amount of fish at the grocery store and see what they say. In this chapter, we consider life as a resource that people organize and that in turn organizes them.

8.1 INTRODUCTION

Like the earlier chapters on fish and forests, this chapter deals with resources that consist of living beings. In the first set of case studies discussed here, non-human resources derive their value from a recognition of their aliveness. That is, the difference between fish farms and sport fishing or cows raised for beef and moose hunted for sport is that in the case of sport hunting and fishing, the aliveness of the resource is central to their value as resources. In other words, part of the attraction of the activity is that the living things have some agency in the activity: the thrill of the hunt or the strategy involved in locating and catching a live animal are part of the appeal.

The second set of case studies focuses on human beings and presents a kind of mirror image: human resources as abstracted from their agents – that is, conceptually separated from the person whose body is in question. Human blood transfusions are one example. Canadian Blood Services serves as a mediator in the use of blood as a resource: when you give or receive a blood donation, you do not know who is receiving your blood or who donated it, and it does not really matter as long as it is safe and of a compatible blood type. Surrogacy – the process through which women gestate fetuses to which they are not at all genetically related – can be understood in a similar way. The resource in question, in this case uteri, is seen as quite separate from the body of which it is a part, and the aliveness of the body is not the main draw.

8.2 WILD(?)LIFE: NON-HUMAN ANIMALS

In this section we consider the organization of animal resources, but we do so in a way that explores how they might blur some of the lines drawn in previous chapters. In Chapter 3, we showed how marine life has been organized into fish stocks; in Chapter 4, we similarly discussed the organization of arboreal life into timber stocks. Here, we discuss animal (and to a lesser extent plant) life more generally. Rather than focusing only on what by now might be a familiar shift from ecosystem component to resource, we also focus here on the agency that these resources might have and on values that inhere in them by virtue of their aliveness and how that exists in tension with a more inert conception of resources.

8.2.1 Pets and Other Companion Species

One of the problems with the nature–society binary is the assumption that humans exist apart from the non-human world, distinct from all other animal species. Non-human animals are admitted to those realms only insofar as they are resources, already objectified, as meat, for example, or iconically as images for visual consumption. But humans do share dynamic, individualized – even personal – and power-laden (political) relationships with non-human animals. The most obvious of these are *domesticated* animals, particularly everyday pets that people share their homes with, such as dogs or cats. *Domesticated* is in italics here because domestication – training to become responsive to the needs of the other – is a two-way process. Donna Haraway (2003) describes dogs, with whom humans have been cohabiting for thousands of years, as "partners in the crime of human evolution" (p. 5). Like our emphasis in this book on humans as both organizers of and organized by, Haraway emphasizes that dogs have organized human beings as much as humans organize canines: "both species ... shape each other throughout the still ongoing story of co-evolution" (p. 29). *Companion species* is a broad term that includes not only pets and other service animals with which people might have individualized relations, but also animals and plants that they rely on for sustenance (cows, chicken, corn, wheat), and even worms (Bennett, 2010, pp. 95–8) or the microbial inhabitants of their bodies (Fishel, 2017). As humans have organized these beings, consciously or unconsciously, they have become vital to human survival.

The ethical obligations that people have toward non-human animals, including whether or how to incorporate them into political decision making, is a vast area of study to which we can only gesture here. Donaldson and Kymlicka (2013) distinguish between three kinds of non-human animals, to each of which humans owe a distinctive set of obligations. First, according to Donaldson and Kymlicka, domesticated animals, like the companion species discussed by Haraway (2003), should be recognized as fully enmeshed members of human–animal communities. Donaldson and Kymlicka argue that wild animals should be entitled to their own sovereign communities (we discuss these in a later section). Third, and perhaps most problematically, are liminal animals, such as raccoons, that are not domesticated but still inhabit human-occupied spaces. Graeme Wynn (2006) suggests that humans' troubled encounters with these kinds of animals (do they belong in urban environments?) may help to illuminate that wild and civilized are cultural concepts, not pre-given or natural categories. If these conceptual boundaries are "neither obvious to, nor observed by, all creatures" (p. xiii), then this may help to shake humans out of a system of classification and simplification that is ultimately ideological. Similarly, in a discussion of the impacts of green infrastructure programs on such liminal species, Christian Hunold (2019) urges the importance of "experienc[ing these kinds of] animals as genuine city residents" (p. 96). Hunold notes that green infrastructure programs seek to make urban spaces more "natural": "ecologically restored waterways, parkland, backyards, urban farms, community gardens, green roofs, rain gardens, and other greened spaces" (p. 89). Too often, however, non-domesticated animal species that are attracted to these renaturalized environments are still treated as pests or nuisances to be managed and contained. Another example here is the push toward "rewilding," that is, intentionally adding individuals or pairs of endangered or extirpated animals to ecologically sensitive areas to return these areas to their "wild" state. This was done with wolves in Yellowstone National Park in the United States, and it triggered a series of changes to the surrounding ecosystem (for a good, brief video on this, see "How Wolves Change Rivers" [Sustainable Human, 2014], linked in the Pedagogical Resources at the end of this chapter). At the same time, rewilding has been critiqued for promoting a particular vision of the wilderness: because ecosystems are always changing and evolving, which version of the past is being framed as an area's natural state?

8.2.2 Fish and Game: Wildness as Economic Resource

In Chapter 3, we told the story of the ever-increasing industrialization of fisheries. At the same time, this treatment of fish as a scientifically managed, mass-produced resource – perhaps most emblematically in large-scale aquaculture operations – is only one side of the story. The transformation from ecosystem component to resource is rarely (if ever) a straightforward linear one, nor is it ever a complete and wholesale transformation. Although resource thinking is expansive, it also generates resistances and alternatives. Although they are not liminal species in the sense discussed earlier, fish and game animals waver in a different kind of middle ground between domesticated and wild. Their significance as a food source means that they are caught up in relations with humans that are at least sometimes antagonistic, and thus they cannot be afforded sovereignty as Donaldson and Kymlicka (2013) want. Unlike domesticated species such as cows or chickens, however, their value as resources depends in part on the extent to which they remain independent from humans and, by providing an alternative to agricultural production, are seen as connecting people with nature. In short, critically examining human relations with these animals allows one to see some of the problems with the nature–society binary.

Fish are particularly interesting here because some are placed on either side of the binary: aquacultural operations, as the name implies, are attempts to place fish (or, more precisely, fish resources) firmly within society, to be managed as closely and carefully as possible. In sport fishing, however, value is derived from the encounter with, and capture of, fish that are wild, that is, outside of society. Game animals occupy a similar niche: outside of society insofar as they are wild animals, but within society insofar as their populations are often the object of intensive managerial attention (wildlife management). Particularly in the case of rural or remote regions, species for sport fishing and hunting constitute a valuable resource to be organized as an important component of economic development strategies, as we show later. Indeed, some of Canada's most powerful and well-known environmental conservation NGOs, such as Ducks Unlimited Canada, see ecological conservation as instrumental to sustaining hunting (and fishing) practices.

In *States of Nature*, Tina Loo (2006) makes the case that conservation laws were an instrument of colonization because in seeking to conserve wildlife, conservation regimes promoted and enforced a particular set of relations between humans and the non-human world. Rather than seeing humans as embedded within ecosystems, they "promoted the non-consumptive use of wildlife, conserving it for the sporting and viewing pleasure of middle-class Canadians" (p. 40). As William Cronon (1995) noted in the case of the United States (and as discussed in Chapter 7), subscribing to this binary separation of nature and society meant that "making a place for wildlife involved pushing some people out of the way" (Loo, 2006, p. 40).

Thus, ironically, the promulgation of an idealized wilderness as a quintessentially Canadian experience made traditional modes of rural subsistence illegal. In its place was put "an urban and bourgeois sensibility about wildlife" (Loo, 2006, p. 40): "wilderness" environments presented as consumption experiences, carefully cultivated from above and for outsiders. At the same time, the organization of these wilderness environments represented a transformation, not an outright destruction, of rural environments. As such, newly organized resources produced new opportunities for at least some rural residents. The knowledge that had come with being more embedded in a local ecosystem could be used in the newly commodified environment, "redeploying [that knowledge] as guides and outfitters or as rural entertainers ... selling local knowledge as woodcraft" (Loo, 2006, p. 8). Although the transformations that Loo (2006) writes about largely took place in the nineteenth and early twentieth centuries, they continue to resonate into the present. The nineteenth- and early twentieth-century "anti-modernist nostalgia [that] led many well-off urbanites to believe that there was no better guarantee of a 'real' backwoods experience than to go with a 'real' woodsman" (p. 54) lives on in contemporary summer camps. In keeping with the construction of an urban and bourgeois sensibility about wildlife, hunting and fishing have largely if not completely been stripped away and replaced with an environmental ethic that positions campers as capable of leaving no trace. This vision of wilderness as a resource to be (sustainably) consumed through encounter is similarly found in outdoor suppliers such as Mountain Equipment Co-op (now Mountain Equipment Company), which for at least two decades has reliably served as an iconic Canadian brand that can be worn to indicate (or perform) the values of sustainability (Fry & Lousley, 2001; on some problems with green consumerism more generally, see Luke, 1997; and Maniates, 2001). This indicates a further shift from encountering wilderness through a commodified but still locally embedded woodcraft to an encounter with wilderness that is technologically mediated to enable getting in and getting out without leaving a trace of one's presence (Turner, 2002).

A similar dynamic of what we might call *conspicuously sustainable consumption* can be seen in other, more common food procurement practices. Things such as free-range poultry or livestock, heirloom varietals, certified organic produce, farmers' markets, community-supported agriculture, and the "eat local" movement can all be seen as connected to a more sustainable lifestyle. Objections to their purported sustainability could be raised, but what they all share is an (explicit or implicit) criticism of the way that food has been organized as a resource in the dominant food paradigm, and they seek to re-embed food production and consumption in more sustainable circuits or authentic contexts and experiences. In each case, however, one can ask, does this constitute an attempt to reconceptualize what has been turned into resources as ecosystem components, or does it seek to reorganize those resources merely to capture value more efficiently from niche markets?

8.2.3 Parks as Spaces for Wildlife

Loo (2006, p. 4) observes that the term *wildlife* is a relatively recent invention, coming into common usage only in the mid-twentieth century, and typically only including land animals. Terrestrial, non-domesticated animals came to comprise a category that was distinctive enough that it required its own term, and that term had to be capacious enough to include animals that were hunted by humans, those that were a nuisance to humans, and others. What is behind this development? Loo cites two developments that are important for our purposes.

The first is that the state takes over what had previously been "a highly localized, fragmented, and loose set of customary, informal, and private practices carried out by a diverse range of individuals and groups" (Loo, 2006, p. 6). As the state took over, what was coming to be known as wildlife was managed with "a more coordinated, encompassing, systematic, and ultimately more scientific approach" (p. 6). In the terms that we have been using throughout this book, and in step with the evolution of ecology as science and the settler dispossession of Indigenous ontologies, wildlife was increasingly seen and treated as a resource to be organized. This is at odds with proponents of animal rights theory, which Donaldson and Kymlicka (2013) define as a "moral framework that acknowledges animals as the bearers of certain inviolable rights" (p. 4).

The second development is the rationale for wildlife management in the first place. The motivation for management helps to show why the category of wildlife needed to be created. Game animals had been (at least ideally) managed with a view to the long-term sustainability of the resource they provided, whereas nuisance animals were managed with a view to mitigating their impacts on human communities (or protecting the resources these animals threatened). Putting these two different kinds of animals, previously subject to quite different management objectives, into a single conceptual category suggests a different kind of management objective, or a different kind of resource that both these kinds of animals (among others) provided. Wildlife conservation was about "the values that would come from conserving it; chief among them was the capacity for people to connect meaningfully – emotionally – with their own natures and with other people. Learning to live with wildlife was, for many who worked on its behalf, about creating ethical human communities" (Loo, 2006, p. 7).

Here, humans are organizing resources – that is, non-human animals – into different management categories. Humans are being organized, too, for example, into different government departments or engaging in different kinds of activities (e.g., conservation tourism vs. pest management). As the last passage from Loo (2006) makes clear, non-human animals are being used to organize humans in terms of their emotional and ethical development.

8.3 HUMAN RESOURCES

Just as forests become timber, marine life becomes fish stocks, and carbon becomes energy as they are abstracted from their ecological context, we suggest that human bodies – or components of them – become identified as resources, such as blood, organs, genes, and surrogates, when they are abstracted from their whole living bodies. In Canada, these resources are generally organized outside of the commodity economy, but they are organized nonetheless.

Here, we give two examples of "human resources" in action: blood and plasma and surrogacy. These probably seem like unusual things to include in a book on Canadian environmental politics, but we suggest that they are relevant here because they are examples of instances in which a biotic entity is converted into a resource in ways that show the kinds of mutual co-organization that we discuss throughout this book. They also show how the line between human (subject) and resource (object) is sometimes fuzzy and always shifting and that resource thinking is expansive, pushing for new things to turn into resources. The examples that follow, however, are different from those in the previous chapters: whereas previous examples showed the mutual organization between society and resources, the examples here show how the resources in question have been organized, but the reverse is not true. Canadians are not geographically organized around the extraction of bodily resources: there are no blood donor or surrogacy towns. We cautiously suggest that this is the case because, at least as we are writing this book, these resources have not been commodified in Canada. In places where human resources have been commodified, there are some ways in which societies have organized more intensively and extensively around them.

8.3.1 Blood and Plasma

In Canada, blood is a resource. Here, we tell the story of why that is so, and why we are addressing it in a book about environmental politics.

In the 1980s, thousands of Canadians contracted hepatitis and HIV after receiving blood transfusions. The crisis became known as the "tainted blood scandal." At the time, blood donation was managed by Red Cross organizations in each jurisdiction. The federal government of the day appointed Justice Horace Krever to conduct an enquiry into the catastrophe. After 247 days of hearing testimony from 474 individuals, the Krever commission released its report (Krever, 1997). A significant portion of the report focused on assessing how the tragedy occurred and how to compensate the victims; it also made recommendations to prevent such a tragedy from recurring. Many of the recommendations pertained to the development of a new blood distribution system. Before the tragedy, each province and territory ran its own blood system. The Krever report explains why this system was a problem: "The federal government was a lax regulator while miserly provincial governments (responsible for funding the system) were more concerned about costs than safety" (Kondro, 1997, p. 1688).

Canadian Blood Services (CBS) arose from the Krever report's recommendation for a system funded by the provinces but responsible to the federal government. This blood distribution system is, of course, not free to run, but separating the funding from the standard setting has a built-in safety mechanism: "By insulating the operator from financial consideration in this way, the reformed blood system has flexibility in the introduction of aggressive safety measures that might otherwise have encountered opposition from a funding organization concerned about the absence of clear evidence for their efficacy" (K. Wilson, 2007, p. 1388).

Krever also recommended that blood in Canada be considered "a public resource for the benefit of all persons in Canada" (Krever, 2007, vol. 3, p. 1047). Unlike many other resources, though, it is a resource that cannot be bought, nor can money be made by providing it. In Canada, a person may volunteer to give blood. Anyone who needs blood will receive it for free.

What is interesting about this setup is that CBS here acts like a corporation when it comes to resourcification of blood. When a forestry company cuts down trees and sells the timber to a retailer, that timber arrives in the store as, for example, a two-by-four-inch piece of wood. When customers purchase that product, they are buying lumber, not a tree. Although the tree species may be known to customers – they may be interested in buying pine, cedar, or oak, for example – its specific story is not. Those customers do not know what forest the tree came from, what trees grew around it, what animals may have used it as habitat, or what smaller plants it sheltered. In a similar way, once blood is abstracted from its "ecosystem," that is, the body in which it previously circulated, it becomes a standardized resource. Other than blood type – A, B, O, positive or negative – the recipient knows nothing about the donor, and the blood's human ecosystem is largely irrelevant to the exchange. Moreover, blood is conceptualized as a national resource: blood donated or received in New Brunswick or Alberta is processed and labelled in exactly the same way, and recipients have no way of knowing the provenance of the transfusion they are receiving.

In the case of blood, then, Canadian society has organized the resource, but it has not organized Canadians in the ways that other resources described in this book have. There are no towns based on a history of blood donation, no university programs dedicated to training blood donors or recipients, and no sports teams, Heritage Minutes, or movies about blood donation. We speculate that this may be because it is not a commodified resource, and we offer the case of plasma as a counter-example to support this claim.

Blood plasma makes up 55 per cent of blood volume. When separated from blood it is a clear, light yellow colour. As CBS (n.d., "What Is Plasma" section) explains, "Plasma is the protein-rich liquid in blood that helps other blood components circulate throughout your body. It supports your immune system and helps control excessive bleeding, which is why plasma donations are important to help treat bleeding disorders, liver diseases and many types of cancer." Like blood, plasma is considered a national resource, and donors cannot be compensated in Canada. A healthy adult can donate blood every 56 days; that same adult can donate plasma as frequently as every 7 days.

Unlike blood donation, however, plasma donation takes a long time and leaves the donor very fatigued; for these reasons, there is far less plasma donation in Canada than blood donation. But the need is not less, and so 80 per cent of plasma products used in Canada come from the United States. Why is there more plasma donation in the United States than in Canada? Because in the United States, donors are paid for plasma donation. For about an hour and a half of plasma donation, a donor is paid about US$30–$50. Other than an increased supply, the paid plasma market has had other effects that show the ways in which commodified resources have organized people.

In effect, plasma donation in the United States is plasma sales. Because the barriers to participating in the plasma market are low – that is, very few things prevent a person from participating – and because the rewards are relatively high as a per-hour price, there are thousands of plasma centres in the United States. These plasma centres are usually located in poorer areas. To measure this phenomenon, James and Mustard (2004) carried out a study in which they surveyed hundreds of plasma centres in the United States and mapped them onto US census data. They found that "plasma clinics were disproportionately overrepresented in areas characterized by socioeconomic disadvantage, residential mobility [homelessness], and active drug sales throughout the period 1980 to 1995. For all 3 measures of neighborhood circumstance, in all years studied, source plasma clinics were more likely to be located in extremely disadvantaged types of neighborhoods" (p. 1227).

Since James and Mustard (2004) conducted their study, the phenomenon in the United States has expanded, in part because of the medical advances that make plasma even more valuable. Today, a unit of plasma for which the donor is compensated US$30–$50 can be processed and sold for US$500: "[Americans'] blood plasma – which historically has been collected disproportionately in the country's poorest communities – is fueling a multibillion-dollar worldwide industry" (Shaefer & Ochoa, 2018, para. 2). Indeed, as Shaefer and Ochoa (2018) put it, "The industry's business model depends on there being plenty of people who need cash quickly." In fact, plasma donation is a significant source of revenue for the poorest US citizens. (See Edin and Shaefer, 2016; for a video interview on this topic, see Feeney, 2015, in the Pedagogical Resources).

Thinking of plasma as a resource that is part of capital circulation lets us think about it in terms of mutual co-constitution. So how has plasma shaped people and societies, other than in the obvious biological ways? For one thing, it shapes medical advances that shape people's lives. It also shapes identities, cultures, and built environments: in the poor neighbourhoods in which plasma centres are located, frequent donors are known as "plassers," and a variety of structural changes have arisen in response to the industry. For example, bus stops are located outside the donation centres, and the centres themselves are located where they are easily accessed by potential donors.

In Canada, plasma is collected mostly by donation through CBS much like blood – that is, from volunteers. Many provinces, including Ontario, British Columbia, and Quebec, have a ban on paid plasma donation; in some provinces, such as Manitoba and Saskatchewan, it is allowed. Critics of the paid model argue that it flies in the face of the Krever report's recommendations and that it preys on at-risk populations. For example, before it was banned in Ontario, plasma clinics were located next to homeless shelters and methadone clinics. However, as supporters of the paid model point out, most of the plasma used in Canada comes from the (paid) American system anyway, and sourcing within Canada would make the country more self-sufficient (Kingston, 2017). Because provinces are regularly introducing, striking down, and reinstating bans, there has not been the same kind of steady plasser income in Canada as there has been in the United States. Perhaps as a result, the same sort of infrastructural changes have not arisen. In the United States, there are bus routes, daily schedules, and communities clustered around plasma centres; in Canada, the paid plasma centres have popped up in poor neighbourhoods, but none seems to have lasted long enough to drive the kind of systemic change that we could describe as mutual co-organization. This finding suggests that the market economy is an important channel through which the nature–society relationship works. Other channels are still available for society to organize nature and vice versa (state regulation, the transmission and reproduction of culture and ideas, the reshaping of built environments, etc.), but when the market portion of the economic channel is closed – when the value of a resource is not described in monetary terms, which usually means its commodification – then certain transformative possibilities are closed off.

The case of surrogacy reinforces this hypothesis, and it also serves as a reminder that those transformative possibilities are not necessarily positive developments.

8.3.2 Surrogacy

A gestational surrogate is a woman who gestates the baby of another family and acts as a biological carrier for a child with whom she shares no genetic material and that she has no intention of raising. For this to occur, an embryo (i.e., a fertilized egg) is implanted into the uterus of the surrogate, who – if all goes as planned – carries the pregnancy to term, delivers the baby, gives it to the IPs, and from that point onward has no involvement in or legal responsibility for the life of the child. Along with options such as adoption and in vitro fertilization, surrogacy is one of several options available to those who want to have children but are unable to do so biologically. In Canada, parents may not legally pay a surrogate for her services. They can reimburse a surrogate for out-of-pocket expenses related to the pregnancy, such as prenatal vitamins, extra food, maternity clothes, or time off work for medical appointments, but the surrogate cannot be paid a salary or stipend for her role as a gestational carrier. In Canada, then, surrogacy is considered a gift from the surrogate to the IPs.

To be sure, surrogacy organizes the lives of those involved, in ways personal, medical, financial, and familial. However – in Canada at least – volunteer surrogacy has not physically organized society in ways analogous to other resources discussed in previous chapters. However, in places where surrogates are paid, the story is different.

India legalized paid surrogacy in 2002 in a bid to increase revenue from medical tourism. One clinic in particular has drawn international attention as a "surrogate village." In Anand, India, is the infamous Akanksha Infertility Clinic run by Dr. Nanya Patel. According to the clinic's website, more than 1,400 babies have been born through the surrogacy process at the clinic. It is a profitable business, drawing an estimated US$400 million a year to India: the IPs pay the clinic about US$25,000–US$30,000, about US$6,500 of which goes directly to the gestational carrier, who lives at the clinic for the duration of the pregnancy (Bhalla & Tapliyal, 2013). The result is a community of surrogates, living in a hotel-like arrangement, who do the work of gestating a baby with whom they share no genetic material. Patel has been both criticized for running a baby factory and lauded for creating jobs for low-income women and offering hope to couples who would otherwise not be able to have children. Whatever one thinks about the desirability of this arrangement, however, the point for our purposes is that the commodification of gestation has organized some small part of Indian economies and cultures.

In 2015, India put restrictions on the surrogacy business, stipulating, among other things, that only Indian citizens may hire Indian surrogates. However, closing this door to foreigners did not decrease the need, and predominantly Western couples started looking elsewhere for potential gestational carriers in countries with more permissive surrogacy laws, such as Thailand, Ukraine, Georgia, and Kenya. Another such country is Cambodia, which saw a surge in surrogacy rates after 2015, when laws were tightened in neighbouring Thailand. Then, as described in the opening story of this chapter, Cambodia passed its own laws in 2017, tightening the rules on surrogacy, and dozens of pregnant surrogates were detained for human trafficking, released only on the condition that they would raise the child they were carrying.

This whipsaw movement from country to country is made possible by the combination of highly mobile capital – IPs can choose to hire a surrogate anywhere they want – and a resource that is not place dependent: there are women willing to act as surrogates in many parts of the world. In this instance, resourcification is ageographical in a way that speaks to the spatial fix described in Box 2.3. Indeed, the ageographical nature of the resources means that what counts here is the biological and financial unevenness, not geographical distribution – although these are related – and so the resourcification of bodies is not part of the national imaginary in the same way as water, fish, timber, or carbon.

8.4 CHANNELS IN ACTION: ORGANIZING LIFE

The preceding case studies raise important questions about the nature–society binary. For example, who determines where on the spectrum of wild to domesticated animals a particular species may fall, and what are the implications of that decision? What does it mean to say that blood is a national resource? And how do the answers to these questions shape the existence and function of people's institutions, communities, and bodies? Drawing on culture and current events, we grapple with some of these questions here.

We can begin with the example of Canadian coins, most of which include an animal image on their surface: the nickel has a beaver, the quarter has a caribou, the loonie is named for the loon it carries on one side, and the toonie has a polar bear. These animals are all wildlife, and indeed it is difficult to imagine such recognition for domesticated animals, such as a dairy cow or pet goldfish, or for creatures considered to be pests, such as racoons, cockroaches, or urban crows. Inasmuch as animal resources have organized Canadian **economies** and **communities** (e.g., think of the impact of the fur trade in the seventeenth and eighteenth centuries), Canadian **culture** and **government** institutions – in this case, the Bank of Canada – are using images of animals as national symbols in ways that imply their wildness. Part of the work of nation building is about classifying ecosystem components as wild, domestic, endangered, economically valuable, and so on. Of course, this does not happen in complete isolation from international factors, but we underscore here that part of the work of nation building is about organizing live resources and managing them accordingly: protecting, exploiting, containing or even eliminating, or revering them.

Outside of non-human animals, the cases of blood, plasma, and surrogacy are two examples of how parts of the human body are transformed from ecosystem components to resources. Several others follow this trend, including organ donation and genetic information. With respect to the latter, recent technological advances have made it much easier to extract genetic information and then recursively manipulate the environment from which it was extracted. The emergence of businesses such as ancestry.ca and 23 and Me, in which customers provide their genetic information for non-medical or non-therapeutic reasons, shows that genetic information is becoming a resource from which economic value can be extracted. Both ancestry.ca and 23 and Me are very profitable companies. The ethical and political questions raised with earlier technologies – should plasma donations be paid for? How does one draw the line between surrogates' expenses and a salary for surrogacy? – multiply and intensify with genetic technologies. What kinds of genetic manipulations should individuals be able to undertake on themselves? On their children? Should insurance companies have access to (or be able to demand) customers' genetic information and set rates accordingly?

Each of these questions (as well as many others) could be answered in isolation from each other, but our hope is that the framework of organizing nature or turning ecosystem components into resources provides a way to think about them as interconnected questions or as symptomatic expressions of a deeper process. These questions become ethical conundrums and political quagmires because the framework of resource thinking is applied to human beings.

Too often, the proposed answers to these questions reflexively reproduce the nature–society binary, presuming that one set of processes and rules can be established for non-human resources while another is reserved for humans. Rather than seeing the issue as drawing that dividing line between humans and non-humans in the right place and strictly policing that border (the nature–society binary), we propose that the problem instead lies in the dominance of resource thinking, and its unthinking application to non-humans as well as humans, in the first place. William Leiss (2011) describes it as "the project of dominating nature ... which begins with the external world (environment), [and] ends at the neurological tissue inside our heads, where our most intimate thoughts and feelings are generated" (pp. 32–3; see Alonzo, 2020, for a recent example).

As the example of surrogacy shows, it may also be possible to think of **bodies** as environments or even as **built environments**, shaped by the societal demand for certain resources. In this light, we can also look back to previous chapters and think about how the work of other kinds of resource extraction has shaped human bodies: black lung among coal miners, log drivers' agility, and cowboys' gait are all examples of bodies shaped by the resources they extract. Writing about women and girls in Nepal, who spend a significant part of each day procuring water for their household, Yoko Arisaka (2001) writes of the pitchers used to carry the water: "Their particular shape and size affect the body and posture of the woman over many years. As a girl grows up, her bodily being-in-the-world adapts to carrying water and becomes part of who she is" (p. 163).

Bodies provide the labour of capitalism for the extraction of fish stocks, timber, coal, and agricultural products. **Bodies** are also organized into **communities** and can be understood as constituent components (e.g., blood, plasma, organs, uteruses). In addition, bodies can become resources that are useful for building nationalism. That is, the benefit here is not necessarily financial, but nationalistic: branding some bodies as Canadian (in this case) **cultural** icons, televising their feats, and claiming their victories builds the kind of nationalism that is a culturally valuable resource.

The Olympics provide a well-known case in point. Examples of national claims to Olympic bodies include daily national medal counts, patriotic advertisements featuring prominent athletes during televised Olympic events, and the athletes themselves donning visible maple leaf tattoos. More generally, the general conflation of national pride, athletes, and consumerism is a boon to advertisers: the attention generated by the Olympics and its renewed sense of nationalism is used to sell everything from mittens to hamburgers to toothpaste. Iconic Canadian brands such as Tim Hortons, Molson, and Air Canada are especially well positioned to profit from the nationalism borne from Olympic events that turn Canadian bodies into resources for nationalism. Indeed, as Bociurkiw (2011) notes of the 2010 Vancouver Olympics, "The televisual representation of the Olympics allowed for a series of things – bodies, signs, highly desired commodities – to be in constant contact. Ads for McDonald's, Coca-Cola, and CTV became almost indistinguishable from the coverage of sporting events themselves. It was this very contiguity, multiplied on innumerable screens and platforms – a mandatory festive viewing – that allowed an imaginary unified Canada to emerge as never before" (p. 149).

Another example of classifying humans as resources is the case of immigration in Canada. Here, individuals are abstracted from their home environment and are given points based on their presumed ability to contribute to Canadian society and economies. Under the Federal Skilled Worker Program, for instance, individuals can score points that determine a grade out of 100. Points are assigned for language fluency, education, age, prearranged work, and adaptability, with each category having its own points value – for example, applicants can get up to 28 points for language if they are fluent in writing and reading in both official languages and up to 10 points for having an offer of full-time employment in Canada. Anyone with 67 points or more qualifies for the program.

The points system was designed to eliminate the racial biases inherent in origin-based selection (Paquet, 2018). Yet, as Gest (2018) points out, the points system can have unintended consequences, such as reinforcing racialized inequalities in the countries of origin or reinforcing gendered inequalities by devaluing skills typically associated with women and rewarding longer periods of uninterrupted work, which is less common for women in countries lacking paid parental leave policies.

For the purposes of our work here, it is worth underscoring that the purpose of the point system is to abstract individuals from their countries of origin, classify them according to their economic value (more points for job experience; fewer for being of an age too young or old to contribute to the market economy), and insert them into the economic engine of the country. The process – in a failed attempt at reducing bias – follows many of the patterns of other resources, only in this case the resource is a human being.

8.5 SUMMARY AND CONCLUSIONS

This book is premised on the idea that humans and their environments are constantly making and remaking each other and that, in Canada, that process is mediated through a variety of channels, including governments, built environments, bodies, cultures, and so on. Why then, a chapter on life? In short, the answer is that the examples presented in this chapter exemplify the nature–society binary and show how the channels work to constantly mediate the relationship between each side of the divide.

In the case of wildlife, classifying a wolf as wild and a dog as domesticated is blurry territory: both are members of the canine family, and indeed wolves are the genetic ancestors of dogs. Yet, the distinction is critical: Banff National Park, for example, has protocols for protecting wolves within park boundaries. No such protection is available for dogs, although the reverse is true in urban settings. Especially when wildlife species hold iconic status – think moose, beaver, bear, wolf, caribou, loon – they become part of the image of the wilderness (note the prefix *wild* in both words), which is a bedrock of the kinds of Canadian identities harkened to in popular culture – for example, the beer ad described in the opening paragraphs of this book.

In the case of body parts and functions such as blood, plasma, surrogacy, and genetic information, the line is even blurrier. Are any of these things nature? Part of the problem with the nature–society binary is that it assumes nature is somewhere "out there"; blood, plasma, surrogacy, and genetic information challenge this premise because they are not out there in some yet-to-be-discovered wilderness, but inside human bodies. They do, however, share some similarities with other resources: once removed from their ecosystem they become abstract, and their provenance is immaterial to their function; they require both labour and capital to extract – capital in this case in the form of medical technologies – and finally, once abstracted (or, in the case of surrogacy, enacted), they are imbricated with capital circulation, that is, money changes hands to gain access to these resources. The exception to this latter point, of course, is blood in Canada, but even CBS classifies that as a resource.

Our aim in this chapter then is to encourage you to think broadly about what constitutes a resource, to ask you to sit in the uncomfortable messiness of defining nature and society, and to reflect on the relationships among money, power, wildness, and resources.

DISCUSSION QUESTIONS

1. How is life (animal or human) similar to other resources addressed in the book? How is it different?
2. One of the themes of this book is the nature–society binary – that is, the false idea that nature and society are two separate things. In the latter half of this chapter, particular components of human bodies are the nature that is reframed as a resource. What channels make this framing possible, and how?

PEDAGOGICAL RESOURCES

Further Viewing

Feeney, L. (2015, December 27). *Living on $2 a day: Exploring extreme poverty in America*. Public Broadcasting Service. https://www.pbs.org/newshour/nation/poverty

Sustainable Human. (2014, February 13). *How wolves change rivers.* Narr. G. Monbiot. YouTube. https://youtu.be/ysa5OBhXz-Q

Videos available on the Bioeconomies Media Project website:

- Collard, R. (2016, March 23). *Animal traffic: The global trade in exotic pets* [Video]. YouTube. https://youtu.be/LcJ_v9tjYaE
- Dempsey, J. (2016, November 23). *Enterprising nature* [Video]. YouTube. https://youtu.be/kGoxXPRY0V8
- ETC Group. (2014, October 29). *What is synthetic biology? Engineering life and livelihoods* [Video]. YouTube. https://youtu.be/C726wUGLdL4

Further Reading

Dempsey, J. (2016). *Enterprising nature: Economics, markets, and finance in global diversity politics.* Wiley Blackwell.

Moore, J.W. (2015). *Capitalism in the web of life: Ecology and the accumulation of capital.* Verso.

Shukin, N. (2009). *Animal capital: Rendering life in biopolitical times.* University of Minnesota Press.

CHAPTER NINE

Resources: Organized and Organizers

Movie critic Roger Ebert (1997) once famously said, "It's not what a movie is about, it's how it is about it" (para. 7). This book addresses resources, but it is not really about resources. Rather, we use resources to talk about some broader issues: politics, culture, colonialism, economies, societies, and the complex relationships that these factors all have with each other and, most relevant to this book, to the environment – however it is defined. Our hope is that throughout this book we have equipped you with some tools to help you to make sense of current and future environmental (and other) issues.

If you take one point from the book, let it be this: society influences environment, and environment influences society.

The first part of this statement (society influences environment) is obvious and is the focus of many excellent environmental politics books that address the role of humans in creating and perpetuating the climate crisis, unprecedented biodiversity loss, crashes in fish stocks, and other local and global environmental crises. It also includes the role that humans play in transforming landscapes in ways that do not necessarily generate crises or catastrophes. The second part (environment influences society) is less obvious and less documented but, we would argue, still a central part of understanding the relationships between people and place. From settlement patterns to the existence (or lack) of specific government departments to cultural icons, the environment plays a leading role in defining how societies are established and how they function. Without subscribing to the kind of environmental determinism that claims that the features of particular human cultures or individuals in them are caused by their local environment, we maintain that humans cannot exert uninhibited mastery over the non-human world because non-human nature often resists, shapes, or sets boundaries on what human societies can do. Also, both society's influence on environment and environment's influence on society are constant and recurring, not one-and-done events. Human societies influence environments, that changed environment influences society, and then that changed society further changes environments, and so on. This recursive dynamic is at the heart of Donald Worster's (1985) question that we quoted at the end of Chapter 2: "How, in the remaking of nature, do we remake ourselves?" (p. 30).

9.1 CHANNELS IN ACTION

Our answer to his question is that it is through channels that people remake themselves in the remaking of nature. The six channels through which nature is organized into resources – governments, communities, built environments, culture and ideas, economies, and bodies and identities – are like well-worn paths through which materials and ideas flow. Although the distinctions between, for example, governments and economies, or communities and built environments, or culture and ideas (knowledges) and bodies and identities are sometimes useful for analytical purposes, we should remember that these are ultimately artificial distinctions. Whether images of wildlife circulating on Canadian currency is slotted into a channel labelled governments, economies, or culture and ideas does not really matter. What is important is the observation that these representations of living beings are icons that help to make possible a national economy and hence a national community and ultimately Canada itself as Canadians know it.

Seeing the operation of these diverse and interconnected channels also allows us to question some of the assumed distinctions that too frequently structure discussions about environmental issues. One is the nature–society binary, which we discuss in more detail later. Another is the presumed distinction between materiality and discourse.

Although it might sometimes be important to distinguish between material or physical changes to the environment on the one hand and discursive representations of the environment (or changes to it) on the other, it is again worth remembering that those distinctions are themselves made by people for analytical purposes. They do not exist in nature itself, unless one assumes that the brains of *Homo sapiens*, which generate the consciousness necessary for discourse, are somehow not part of the natural world.

A final point to emphasize about channels here is that not only do the flows of materials and discourses connect people to places and vice versa, they also reshape people and places. People, both individually and collectively, are constantly evolving as they try to survive and thrive in the places in which they find themselves. And places – at multiple scales – are constantly evolving as they are remade by human and non-human inhabitants, or anthropogenic and non-anthropogenic forces.

9.2 COMMON THEMES

Despite the temporal, geographic, and material differences between each case study, the themes are the same: commodification, Indigenous dispossession, and promotion of the nature–society binary. Moving forward, these themes can constitute your "organizing nature" glasses: the lenses you can look through to see and understand the world in a nuanced and interconnected way. Together, these three themes are what we call resource thinking: a particular mindset that, consciously or not, sees ecosystem components as things that are separate from people and as things from which profit can be derived.

9.2.1 Commodification

One of this book's themes is ecosystem commodification – assigning monetary and exchange value to particular ecosystem components. Commodification is the process through which some ecosystem components can be removed from their ecological milieu and sold for profit. Resource thinking makes a distinction between something being useful for subsistence and something being commodified: it is the difference between fish being harvested for subsistence or cultural value and fish harvested and sold at a commercial scale, or the difference between using wood to build a dwelling and clear-cutting to export wood around the world. For fish to become fish stocks, or wood to become timber, or land to become real estate, both the seller and the purchaser must agree that this thing has financial value, in some sense separate from its immediate use.

In enabling the transition from ecosystem component to resource, the process of commodification not only extracts ecosystem components from their environments, it also abstracts them: When you buy a can of salmon, you might know what species of salmon it is and what country it comes from, but you do not know what stream it was born in, where it travelled to, or any other information that embeds salmon into a particular web of ecological and social relationships. Resourcification erases that ecological history.

Over time, resource thinking becomes self-reinforcing. Take the transition from forest to lumber as an example: extracting and abstracting trees to sell on a commodity market enables and encourages the planting of profitable trees as crops – pine, for example – which in turn creates a new kind of managed forest whose ecosystem functions are secondary to its role as a cash crop. Those trees are then very clearly resources and are managed as such. Or take agricultural land as another example. Once the land has been cleared of brush and sold, its value lies in its properties as they relate to farming – things such as soil composition, grade and drainage, and climate. Reframing the land in this way reinforces its value as a resource rather than, say, part of a migratory path or a habitat for pollinators. Recently, attempts to recognize those important functions have taken the form of payment for ecosystem services (see Chapter 4). However, we join others (e.g., Dempsey, 2014) in suggesting that such mechanisms reinforce, rather than challenge, resource thinking: they commodify the land, just through different mechanisms.

9.2.2 Indigenous Dispossession

Resource thinking comes at the expense of Indigenous Peoples, who are dispossessed of both land and culture through the processes associated with resource thinking. Dispossession is not a historical event that happened hundreds of years ago and is over; it is an ongoing process that is baked into language, property ownership, legal systems, economies, and nearly every other dimension of the Canadian cultural landscape. In many instances, resource extraction happened, and continues to happen, in violation of treaty rights, which can cause irreparable damage to landscapes with cultural importance and have negative effects disproportionately experienced by Indigenous communities. Although resource extraction may provide economic opportunity in some of the remote regions where reserves have been established, it is the settler colonial system that "reserved" Indigenous communities to those remote regions in the first place. Moreover, resource thinking, which sees the world in terms of disconnectable, extractable, exchangeable, individual things, is at odds with world views that emphasize interconnectedness and interrelatedness. Indeed, seeing the planet as a site of extraction is, itself, a form of dispossession (A. Cohen et al., 2021).

It is important to note that it is not as though Indigenous Peoples were entirely dispossessed of land and culture, and now the story is over. Far from it: Indigenous Peoples across the country are actively engaged in land management and protection across the country. From the Centre for Indigenous Environmental Resources in Winnipeg (https://yourcier.org), to Indigenous legal scholars fighting in the courts, to brilliant thinkers like Audra Simpson (2014) and Glen Coulthard (2014), to Indigenous education programs at the Dechinta Centre for Teaching and Learning (https://www.dechinta.ca), to the establishment of the Indigenous Guardians Network – the "eyes and ears" on traditional territories (https://www.ilinationhood.ca) – there are so many examples of the ways in which Indigenous Peoples and their allies are pushing back against historical and contemporary colonialism, of which resource thinking is a central part.

9.2.3 Artificial Nature–Society Binary

A third theme of the book is the widespread perpetuation of a nature–society binary, which is the idea that nature and society are two entirely separate things. We hope that this book has shown you many ways in which this is not at all the case, but, nevertheless, the idea that nature is there and society is here is a pervasive one: once you see a dividing line between nature and society, it is difficult to unsee it. This way of thinking is both incorrect and harmful – if nature is out there, separate from society, then it is easier not to worry about polluted water, climate change, or myriad other issues. Obviously that is not the case, and one aim of the book is to help break down collective mental barriers between people and place.

The nature–society binary is equally problematic in environmentalist spheres, where the idea of saving the environment, with the implication that the environment is a place where humans are absent (e.g., wilderness), resonates with those who want to make a difference. The binary is not only an individual mental framing; it also structures Canadians' collective political institutions. For example, most provincial and territorial governments – and the federal government – have departments of health, departments of environment, and departments of natural resources, each of which operates as its own unit. This division means that drinking water is managed under an entirely different system than environmental water, but of course it is all the same water moving through the water cycle. Similarly, pulp and paper processing can affect air quality, but departments of natural resources and departments of health are two separate entities. Some companies and organizations have a sustainability division, but it often operates in parallel to the rest of the organization rather than being baked into the function of the whole organization. For this reason, the persistence of the binary can also seriously weaken calls to environmental action, because those actions become focused on the environment rather than on human health and survival.

Early in the book we introduced a paradox: that Canadians are fiercely proud of Canadian nature and want to preserve it while also depending economically on its extraction and commodification. We suggested that the paradox exists because of the nature–society binary. The ideas that parks or wilderness areas are separate from other land, that forests are separate from lumber, that drinking water is separate from other water (as in the division of responsibilities for water), or that commercially valuable fish are separate from the waters they inhabit (as in the *Fisheries Act*) are all examples of the nature–society binary in action. Imagining forests as entirely separate from lumber, for example, enables a person to hold two truths: that forests are a wilderness to be protected and that lumber is a resource to be extracted. Understanding when, how, and especially why this binary is promoted – and the paradox it enables – is central to understanding environmental politics. It is this last point – the why – that we turn to next.

9.3 WHY DOES RESOURCE THINKING MATTER?

The three dimensions of resource thinking – commodification, dispossession, and the artificial nature–society binary – have real-world consequences for real-world places and people. Indeed, resource thinking is so embedded in settler culture that it is often invisible, but this way of thinking – like all world views – makes a difference in the kinds of lives that people, and the other beings that they share environments with, can live. Recalling a few examples from past chapters will help to show what we mean.

In Chapter 4, we talked about two ways in which forests are commodified: either being cut down (i.e., timber) or not being cut down (i.e., tourism, carbon credits). Is there a way to envision a future for sustainable forests that does not involve some form of commodification? In Chapter 5, we noted former Environment Minister Stéphane Dion's statement that "there is no minister of the environment on Earth who can stop [the oilsands] from going forward, because there is too much money in it" (as quoted in Haley, 2011, p. 97). And yet, of course, for millennia, that bitumen did stay in the ground, untouched. And on it goes. So when you hear or read statements like "Canada has a resource-based economy" or "energy resources from northern Alberta need to get to market," we encourage you to step back, put on your "organizing nature" glasses, and see that all these questions are focused on the most profitable, equitable, or efficient way of extracting or processing ecosystem components. These kinds of statements see – and to a large extent do not even question – the possibility and desirability of comprehending the world as a set of resources that can and should be organized for maximum gain (again, with differences in how gain is to be understood and measured). The prevalence of this way of seeing is less surprising if we recall the point noted earlier that Canada as a national community is built on resource thinking. Insofar as political parties are entities that are organized to seek power through the state (being elected to form a government), they are necessarily invested in this national project, so, with the partial exception of the Green Party, this view is rarely questioned in party politics. Seeing resource thinking as something constructed and contestable, rather than as natural and inevitable, allows one to see how conventional framings tilt the table to create winners and losers.

9.3.1 Winning and Losing

Who wins and loses from resource thinking? Private companies and their owners are one example of a group that benefits from resource thinking. In particular, companies that are engaged in what are often described as extractive industries could not operate without resource thinking. As we have emphasized, resource extraction depends on the abstraction inherent in resource thinking. It takes a resource-thinking frame of mind to look at a forest and see trees that can be cut down and processed into timber to be sold to customers an ocean away. More generally, businesses are designed to earn profits for their owners (in the case of publicly traded corporations, they have a fiduciary obligation to their shareholders). Profits can only be earned if the company has something that it can sell, and something can only be sold (its ownership transferred) if it is conceptualized as a commodity – something that can be exchanged or removed from its (ecological) context – or, in other words, if it is organized as a resource. Of course, individual companies pay workers to plant trees or raise animals, and they pay governments for licences to harvest, but the ecosystemic processes by which those things grow are free.

Whether the employees of those companies benefit from the deployment of resource thinking is more complicated because the where, when, and how of resource thinking matters. Millions of Canadians have jobs that are based on resource extraction and processing. For example, as seen in Chapter 5, the high-paying jobs in oil sands extraction afford many workers a relatively high level of material consumption and comfort. Extractive jobs can also provide a degree of psychological security and identity, whether that is in the form of fishing culture (see Chapter 3) or petro-masculinity (see Chapter 5). At the same time, we have also provided examples in which the organization of extractive industries has been detrimental to workers: the economic deprivation and bodily toll of nineteenth-century forestry (Chapter 4) or coal mining (Chapter 5) or the twenty-first century commodification of surrogacy (Chapter 8).

Previous chapters have also discussed geographic communities whose growth or even creation was made possible by resource thinking. The most obvious examples are the resource towns with all the spin-off jobs from resource extraction: schools, hospitals, restaurants, and so on. However, resource town residents may live with poor air or water quality, live in an unstable climate, or have to deal with other environmental hazards. Moreover, resources that are extracted in one locale are often organized to enable their flow elsewhere, a process that in global terms has been described as unequal ecological exchange (Hornborg, 1998). Resources are accumulated, often far from the site of extraction, enabling the concentration of wealth and power (both public and private) in urban centres. In this sense, it is not only places such as fishing outports or forestry and mining towns but also Canada's biggest cities that have been built by the organization of nature and its transformation into resources. Indeed, Canada as a whole is a country built through the organization of resources.

Indigenous Peoples are perhaps the most obvious example of those who have been negatively affected by resource thinking. The settler colonial dispossession on which Canada was founded has been both physical and cultural. Physically, resource thinking has enabled the physical occupation of the landscape by settlers, often to the exclusion of its original inhabitants. Indigenous lands are confined to marginal and remote reserves, often deprived of basic amenities such as clean drinking water and adequate housing. Culturally, the more holistic understandings of land characteristic of Indigenous world views (discussed in Chapter 7) are replaced by a conception of land as property. This change in cultural understanding was and is enforced (at times, violently) through a range of institutions, from the residential school system that sought to assimilate Indigenous children into settler Canadian culture to legal regimes that assert the right (jurisdiction) of governments to approve pipelines or other development projects through traditional Indigenous territories.

More generally, those who are disadvantaged by resource thinking are those whose voices are diminished or lacking in existing political and economic structures. The reproductive work that is largely done by women is not recognized in the formal (market) economy and is thus devalued by resource thinking. As we showed in Chapter 2, the environmental justice movement arose in response to the inequitable treatment of racialized minority communities when it came to the distribution of environmental harms and hazards. Youth climate strikes have drawn attention to the lack of voice that is afforded future generations in political institutions. Environmentalists, among others, have long sought to bring the concerns and interests of the non-human world into political decision making. To the extent that resource thinking is the dominant way of thinking about (and interacting with) the world, all of these groups struggle to be recognized and have their vital needs met.

9.3.2 Why Is It Important to Think beyond Policy?

Conversations about resource management often pivot on questions of policies – making them, enacting them, evaluating them, and so on. We hope this book has shown you, though, that policies are only part of the story. Policies do not exist in a vacuum; rather, they exist in a matrix of cultural, social, economic, and political features that shape, and are shaped by, the physical landscape on which Canada is situated. Understanding that, for example, discussions about water exports are not just about the water helps to make sense of the current political landscape.

Finally, thinking beyond policy also creates room to consider not only how humans shape the landscape, but how it has shaped them. In the Molson beer ad described at the start of this book, the voiceover asks, "Why are we the way we are?... It's this land that shapes us." Yes, it is easy to make light of a beer ad and the over-the-top narration that accompanies it, but Molson makes a good point here (even if it was not intentional): the landscape does shape Canada. It shapes its institutions, its economy, where Canadians settle, what they learn, their art, what money looks like, and other icons big and small. Things such as forestry degrees, fishing licences, grain elevators, and national parks are all examples of objects and institutions that were built specifically to enable the extraction of value from particular landscape components. In the creation of these things, ecosystems have played a role in shaping the Canadian landscape – both physical and institutional – and highlight the point that people and place are in a mutually constitutive relationship. Focusing on policy alone only shows one side of this relationship and tells half a story.

Glossary

Anthropocene See Box 2.2.

Anthropocentric A viewpoint in which human activity is at the centre. In this view, resources are defined by their usefulness to humans.

Biodiversity The variety of species of life, either within a particular area or globally.

Bitumen A thick, tarlike fossil fuel found in sandy deposits around the world, including in northern Alberta. Once refined, it is used in a variety of energy applications, such as gasoline.

Boil-water advisories Health warnings issued when drinking water from a municipal supply contains levels of bacteria harmful to human health. Boiling the water kills the bacteria. Note that boil-water advisories do not resolve issues with heavy metals or other toxins in water.

Capital The accumulated product of past labour, invested for further production. Capital can take a fixed form (e.g., building a factory) or a mobile form (e.g., cash or other financial assets).

Capitalism A dynamic economic system, developed over the past few hundred years, in which most decisions about what is produced and how are made by private actors (individuals or corporations), and most people earn their living by selling their labour (wages or salaries).

Commodification The process of assigning monetary value to a thing, making it available to be bought or sold or both.

Crown, the The symbolic representation of the state in Canada.

Crown corporation A corporation that is publicly owned and operated. Several power companies in Canada are Crown corporations. At the federal level, organizations such as VIA Rail and Canada Post are examples of Crown corporations.

Discourse (or discursive) The language, both written and spoken, used to describe particular places, people, or processes. *Discursive analysis* is the examination of how something is described.

Ecosystem component A specific environmental item or process. In resource thinking, ecosystem components are reframed as resources.

Ecosystem services Environmental features or functions that are useful to humans, such as shade, pollination, or water filtration.

Environmental determinism The idea that the characteristics of an individual or community are reflective of the features of their environment.

Environmental footprint A symbolic representation of environmental impact. Usually measured in square kilometres or miles, it shows the area needed to grow or produce goods that support the life(style) of a person or group.

Environmental justice See Box 2.4.

Fracking A process for oil and gas extraction that uses drilling and high-pressure injection of liquids to force open fissures.

Heritage Minutes A series of 60-second vignettes produced by Historica Canada. Heritage Minutes air or aired on television, in movie theatres, and online and depict key moments in Canadian history for education and entertainment.

Indigenous Peoples who lived in a territory before colonization. In Canada, Indigenous Peoples include First Nations, Métis, and Inuit.

Institutions Permanent solutions to permanent problems.

Jurisdiction The power to make legally binding decisions.

Legibility The processes through which highly complex situations are simplified for them to make sense, usually to a central authority, such as the state.

Nation A group of people united by a sense of common belonging, usually associated with a particular territory.

Nature–society binary The idea that the environment (i.e., nature) and people (i.e., society) are two separate things.

Ontology Knowledge about the nature of reality, being, and existence.

Resource thinking An anthropocentric way of thinking wherein some parts of the ecosystem are understood as resources. Resource thinking reinforces commodification, Indigenous dispossession, and the nature–society binary.

Rights Entitlements or legal claims that can be made against either the state or other people.

Settler colonialism See Box 7.1.

Staples Goods that are important in consumers' everyday lives. A *staples economy* is one that is organized around the extraction or production of staples for export. The *staples trap* is when an economy is unable to adjust to changes in demand for the staples it produces.

States Institutions that have a monopoly on the legitimate use of force within a particular territory. States persist even while governments (formed by specific political parties) change.

Sustainability The ability of something (e.g., a stock of resources) to be maintained over time. Initially used by environmentalists to refer to environmental resources or amenities, it is increasingly referred to as having three dimensions: environmental, social, and economic.

Terra nullius A Latin phrase literally meaning "nobody's land"; a legal doctrine used to justify European colonization.

Wilderness Nature devoid of humans. A problematic idea because the idea of wilderness usually erases Indigenous presence and reinforces the nature–society binary.

References

Agriculture Canada. (2017). *An overview of the Canadian agriculture and agri-food system 2017* (AAFC No. 12714E). Agriculture and Agri-Food Canada. https://publications.gc.ca/collections/collection_2018/aac-aafc/A38-1-1-2017-eng.pdf

Agyeman, J., Cole, P., Haluza-DeLay, R., & O'Riley, P. (Eds.). (2010). *Speaking for ourselves: Environmental justice in Canada*. UBC Press.

Aït-Ouyahia, M. (2006). Is there a place for public-private partnerships for municipal drinking water in Canada? *Horizons, 9*(1), 45–8.

Alberta. (2017, March 28). *Demographic spotlight: Interprovincial employees in Alberta: Industrial profile by major region of origin*. https://open.alberta.ca/dataset/0923631a-1ee7-4621-b048-93a99fe2d07b/resource/e114d08a-b08a-434a-a398-f7258b0e248c/download/2017-0328-interprovincial-employees-in-alberta-industrial-profile.pdf

Alberta Water Portal. (2012, July 30). *Agriculture and irrigation in Alberta*. https://albertawater.com/what-is-water-used-for-in-alberta/agriculture-in-alberta

Albrecht, G., Sartore, G.M., Connor, L., Higginbotham, N., Freeman, S., Kelly, B., Stain, H., Tonna, A., & Pollard, G. (2007). Solastalgia: The distress caused by environmental change. *Australasian Psychiatry, 15*(Suppl. 1), S95–8. https://doi.org/10.1080/10398560701701288

Alonzo, I. (2020, August 4). *Elon Musk's Neuralink chip will soon allow users to take charge of moods and emotions*. Tech Times. https://www.techtimes.com/articles/251574/20200804/elon-musk-neuralink-update-mood-control.htm

Anderson, B. (1983). *Imagined communities: Reflections on the origin and spread of nationalism*. Verso.

Andrew-Gee, E. (2020, March 3). "The railways got very wealthy on our land": How rail's colonial past made it a target for blockades. *The Globe and Mail*. https://www.theglobeandmail.com/canada/article-the-railways-got-very-wealthy-on-our-land-how-rails-colonial-past/

Arisaka, Y. (2001). Women carrying water: At the crossroads of technology and critical theory. In W.S. Wilkerson & J. Paris (Eds.), *New critical theory: Essays on liberation* (pp. 155–74). Rowman & Littlefield.

Arnold, S. (2008). Nelvana of the north, traditional knowledge, and the northern dimension of Canadian foreign policy. *Canadian Foreign Policy Journal, 14*(2), 95–107. https://doi.org/10.1080/11926422.2008.9673465

Assembly of First Nations. (2018, May 18). *First Nations fishing rights – fact sheet*. https://www.afn.ca/first-nations-fishing-rights-fact-sheet/

Atherton, E., Risk, D., Fougere, C., Lavoie, M., Marshall, A., Werring, J., Williams, J.P., & Minions, C. (2017). Mobile measurement of methane emissions from natural gas developments in northeastern British Columbia, Canada. *Atmospheric Chemistry and Physics, 17*(20), 12405–20. https://doi.org/10.5194/acp-17-12405-2017

Auld, A. (2001, May 21). Historic Cape Breton coal mining industry unravels at the seams. *The Globe and Mail*. https://www.theglobeandmail.com/news/national/historic-cape-breton-coal-mining-industry-unravels-at-the-seams/article4148030/

Bakker, K. (Ed.). (2006). *Eau Canada: The future of Canada's water*. UBC Press.

Bakker, K. (2010). *Privatizing water*. Cornell University Press.

Baldwin, A. (2009). Ethnoscaping Canada's boreal forest: Liberal whiteness and its disaffiliation from colonial space. *The Canadian Geographer/Le Géographe canadien, 53*(4), 427–43. https://doi.org/10.1111/j.1541-0064.2009.00260.x

Barry, J. (2012). *The politics of actually existing unsustainability: Human flourishing in a climate-changed, carbon constrained world*. Oxford University Press.

Bavington, D. (2010). From hunting fish to managing populations: Fisheries science and the destruction of Newfoundland cod fisheries. *Science as Culture, 19*(4), 509–28. https://doi.org/10.1080/09505431.2010.519615

Baxter, J. (2019, May 6). Fracking is back on the agenda in Nova Scotia. *Halifax Examiner*. https://www.halifaxexaminer.ca/province-house/fracking-is-back-on-the-agenda-in-nova-scotia/

Benidickson, J. (2010). Cleaning up after the Log Drivers' Waltz: Finding the Ottawa River watershed. *Les Cahiers de droit, 51*(3–4), 729–48. https://doi.org/10.7202/045731ar

Bennett, J. (2010). *Vibrant matter: A political ecology of things*. Duke University Press.

Berger, P.L., & Luckmann, T. (1967). *The social construction of reality: A treatise in the sociology of knowledge*. Doubleday.

Berkes, F., Colding J., & Folke, C. (2000). Rediscovery of traditional ecological knowledge as adaptive management. *Ecological Applications, 10*(5), 1251–62. https://doi.org/10.1890/1051-0761(2000)010[1251:ROTEKA]2.0.CO;2

Berton, P. (1971). *The last spike: The great railway 1881–1885*. McClelland & Stewart.

Bhalla, N., & Tapliyal, M. (2013, September 30). *As surrogacy industry booms; India seeks controls*. NBC News. https://www.nbcnews.com/businessmain/surrogacy-industry-booms-india-seeks-controls-8C11300035

Biro, A. (2002). Wet dreams: Ideology and the debates over Canadian water exports. *Capitalism Nature Socialism, 13*(4), 29–52. https://doi.org/10.1080/10455750208565499

Biro, A. (2019). Reading a water menu. *Journal of Consumer Culture, 19*(2), 231–51. https://doi.org/10.1177/1469540517717779

Blackmore, O. (2019, May 13). *Defining a "moderate livelihood": Part 1*. Ku'ku'kwes News. https://kukukwes.com/2019/05/13/defining-a-moderate-livelihood-part-1/

Bluhdörn, I. (2007). Sustaining the unsustainable: Symbolic politics and the politics of simulation. *Environmental Politics, 16*(2), 251–75. https://doi.org/10.1080/09644010701211759

Bociurkiw, M. (2011). *Feeling Canadian: Television, nationalism, and affect*. Wilfrid Laurier University Press.

Boyd, D.R. (2003). *Unnatural law: Rethinking Canadian environmental law and policy*. UBC Press.

Boyd, D.R. (2011). No taps, no toilets: First Nations and the constitutional right to water in Canada. *McGill Law Journal, 57*(1), 81–134. https://doi.org/10.7202/1006419ar

Boyd, D.R. (2017). *The rights of nature: A legal revolution that could save the world*. ECW Press.

Bozikovic, A. (2017, October 17). Google's Sidewalk Labs signs deal for "smart city" makeover of Toronto's waterfront. *The Globe and Mail*. https://www.theglobeandmail.com/news/toronto/google-sidewalk-toronto-waterfront/article36612387/

Braidotti, R. (2013). *The posthuman*. Wiley.

Brei, V.A. (2018). How is a bottled water market created? *WIREs Water,* 5(1), e1220. https://doi.org/10.1002/wat2.1220

Broadcasting Act, SC 1991, c. 11, s. 3. https://laws-lois.justice.gc.ca/eng/Const/page-4.html#h-17

Brown, D. (2005). *Salmon wars: The battle for the West Coast Salmon Fishery*. Harbour Publishing.

Brown, S. (2019, June 5). *New Brunswick Indigenous chiefs left "blindsided" by decision to lift fracking moratorium*. Global News. https://globalnews.ca/news/5356115/indigenous-chiefs-issue-warning-gas-fracking/

Bueckert, K. (2016, August 26). *Township of Centre Wellington tried to buy Elora well now owned by Nestle*. CBC News. https://www.cbc.ca/news/canada/kitchener-waterloo/middlebrook-well-elora-nestle-centre-wellington-purchase-1.3735854

Bullard, R.D. (2000). *Dumping in Dixie: Race, class, and environmental quality* (3rd ed.). Westview. (Original work published 1990)

Bullard, R.D., & Wright, B. (2012). *The wrong complexion for protection: How the government response to disaster endangers African American communities*. NYU Press.

Butler, J. (1993). *Bodies that matter: On the discursive limits of "sex."* Routledge.

Canada Energy Regulator. (2021). *Canada's pipeline system 2021: Economics of CER-regulated infrastructure*. Retrieved February 15, 2023 from https://www.cer-rec.gc.ca/en/data-analysis/facilities-we-regulate/canadas-pipeline-system/2021/introduction.html

Canadian Blood Services. (n.d.). *Plasma for life*. Retrieved October 17, 2019 from https://blood.ca/en/plasma

Canadian Charter of Rights and Freedoms, Part 1 of the *Constitution Act, 1982*, being Schedule B to the *Canada Act 1982* (UK), 1982, c. 11.

Canadian Pacific. (n.d.). *Our history*. Retrieved February 20, 2023 from https://cpconnectingcanada.ca/our-history/

Cannavò, P.F. (2007). *The working landscape: Founding, preservation, and the politics of place*. MIT Press.

Carrington, D. (2016, August 29). The Anthropocene epoch: Scientists declare the dawn of human-influenced age. *The Guardian*. https://www.theguardian.com/environment/2016/aug/29/declare-anthropocene-epoch-experts-urge-geological-congress-human-impact-earth

Carter, A.V. (2016). Environmental policy and politics: The case of oil. In D.L. VanNijnatten (Ed.), *Canadian environmental policy and politics: The challenges of austerity and ambivalence* (4th ed., pp. 292–306). Oxford University Press.

CBC News. (2001, May 3). *Environmentalist begins Parliament Hill hunger strike*. CBC. https://www.cbc.ca/news/canada/environmentalist-begins-parliament-hill-hunger-strike-1.284072

CBC News. (2010, June 25). *Toxic timeline*. https://www.cbc.ca/news/canada/toxic-timeline-1.888186

CBC News. (2012, October 31). *No "smoking gun" for Fraser River sockeye salmon collapse*. https://www.cbc.ca/news/canada/british-columbia/no-smoking-gun-for-fraser-river-sockeye-salmon-collapse-1.1256706

CBC News. (2017, June 8). *Ontario confirms bottled water companies to pay more as of August 1*. https://www.cbc.ca/news/canada/kitchener-waterloo/ontario-bottled-water-taking-fee-going-up-1.4151685

CBC Radio. (2015, April 22). *Commemorating 25th anniversary of historic Odeyak voyage*. CBC. https://www.cbc.ca/radio/asithappens/as-it

-happens-Wednesday-edition-1.3044034/commemorating-25th-anniversary-of-historic-odeyak-voyage-1.3044498
Chong, E., & Whewell, T. (2018). *Paid to carry a stranger's baby – Then forced to raise it*. BBC News. https://www.bbc.co.uk/news/resources/idt-sh/surrogates
Civil Marriage Act, SC 2005, c. 33. https://laws-lois.justice.gc.ca/eng/acts/c-31.5/page-1.html
Coates, T. (2014, June). The case for reparations. *The Atlantic Monthly*. https://www.theatlantic.com/magazine/archive/2014/06/the-case-for-reparations/361631/
Cohen, A., Matthew, M., Neville, K., & Wrightson, K. (2021). Colonialism in community-based monitoring: Knowledge systems, finance, and power in Canada. *Annals of the Association of American Geographers, 111*(7), 1988–2004. https://doi.org/10.1080/24694452.2021.1874865
Cohen, B.I. (2012, October). *The uncertain future of Fraser River sockeye: Causes of the decline* (Vol. 2; Catalogue No. CP32-93/2012E-2 [v. 2]). Commission of Inquiry Into the Decline of Sockeye Salmon in the Fraser River. http://publications.gc.ca/collections/collection_2012/bcp-pco/CP32-93-2012-2-eng.pdf
Cole, D. (1978). Artists, patrons and public: An enquiry into the success of the Group of Seven. *Journal of Canadian Studies/Revue d'études canadiennes, 13*(2), 69–78. https://doi.org/10.3138/jcs.13.2.69
Commoner, B. (2020). *The closing circle: Nature, man, and technology*. Dover Publications. (Original work published 1971)
Conca, K. (2005). *Governing water: Contentious transnational politics and global institution building*. MIT Press.
Conservation International. (2019). *Nature now* [Video]. https://www.conservation.org/video/nature-now-video-with-greta-thunberg
Constitution Act, 1867, 30 & 31 Victoria, c. 3 (UK). https://laws-lois.justice.gc.ca/eng/const/page-1.html
Constitution Act, 1982, being Schedule B to the *Canada Act 1982* (UK), 1982, c. 11. https://laws-lois.justice.gc.ca/eng/const/page-12.html#h-39
Convention on the International Trade in Endangered Species of Wild Fauna and Flora. (1973). https://cites.org/eng/disc/text.php
Coole, D., & Frost, S. (Eds.). (2010). *Introducing the new materialisms*. Duke University Press.
Costanza, R., d'Arge, R., de Groot, R., Farber, S., Grasso, M., Hannon, B., Limburg, K., Naeem, S., O'Neill, R.V., Paruelo, J., Raskin, R.G., Sutton, P., & van den Belt, M. (1997). The value of the world's ecosystem services and natural capital. *Nature, 387*(6630), 253–60. https://doi.org/10.1038/387253a0
Coulthard, G.S. (2014). *Red skin, White masks: Rejecting the colonial politics of recognition*. University of Minnesota Press.

Council of Canadians. (2020). The new NAFTA: How it impacts our water. https://canadians.org/wp-content/uploads/2022/06/factsheet-nafta-water.pdf

Cox, K.R. (1996). The difference that scale makes. *Political Geography, 15*(8), 667–9. https://doi.org/10.1016/0962-6298(96)82552-2

Cronon, W. (1983). *Changes in the land: Indians, colonists, and the ecology of New England*. Hill & Wang.

Cronon, W. (1995). The trouble with wilderness; or, getting back to the wrong nature. In W. Cronon (Ed.), *Uncommon ground: Rethinking the human place in nature* (pp. 69–90). W.W. Norton.

Crutzen, P. (2006). The Anthropocene. In E. Ehlers & T. Krafft (Eds.), *Earth system science in the Anthropocene* (pp. 13–18). Springer.

Cunsolo, A. & Ellis, N.R. (2018). Ecological grief as a mental health response to climate change-related loss. *Nature Climate Change, 8*, 275–81. https://doi.org/10.1038/s41558-018-0092-2

Daggett, C. (2018). Petro-masculinity: Fossil fuels and authoritarian desire. *Millennium: Journal of International Studies, 47*(1), 25–44. https://doi.org/10.1177/0305829818775817

Dale, S. (1999). *Lost in the suburbs: A political travelogue*. Stoddart.

Delanty, G., & Mota, A. (2017). Governing the Anthropocene: Agency, governance, knowledge. *European Journal of Social Theory, 20*(1), 9–38. https://doi.org/10.1177/1368431016668535

Dempsey, J. (2014). *Enterprising nature: Economics, markets, and finance in global biodiversity politics*. Wiley-Blackwell.

Deneault, A. (2019, April). The Irvings, Canada's robber barons. *Le Monde Diplomatique*. https://mondediplo.com/2019/04/13canada

Denevan, W.M. (1992). The pristine myth: The landscape of the Americas in 1492. *Annals of the Association of American Geographers, 82*(3), 369–85. https://doi.org/10.1111/j.1467-8306.1992.tb01965.x

Department of Fisheries and Oceans. (2018). *Canada's fisheries: Fast facts 2017*. Economic Analysis and Statistics. https://waves-vagues.dfo-mpo.gc.ca/library-bibliotheque/40706990.pdf

Department of Fisheries and Oceans. (2022). *Mandate and role*. Retrieved February 21, 2023 from https://www.dfo-mpo.gc.ca/about-notre-sujet/mandate-mandat-eng.htm

Department of Natural Resources and Renewables. (n.d.). *Nova Scotia's historic underground coal mine workings information*. Government of Nova Scotia. Retrieved July 5, 2019 from https://novascotia.ca/natr/meb/hazard-assessment/historic-coal-mine-workings.asp

Donaldson, S., & Kymlicka, W. (2013). *Zoopolis: A political theory of animal rights*. Oxford University Press.

Dunn, G., Bakker, K., & Harris, L. (2014). Drinking water quality guidelines across Canadian provinces and territories: Jurisdictional variation in the context of decentralized water governance. *International Journal of*

Environmental Research and Public Health, 11(5), 4634–51. https://doi.org/10.3390/ijerph110504634

Dussault, R., Erasmus, G., Chartrand, P., Meekison, J., Robinson, V., Sillett, M., & Wilson, B. (1996). Report of the Royal Commission on Aboriginal Peoples (5 vols.). https://www.bac-lac.gc.ca/eng/discover/aboriginal-heritage/royal-commission-aboriginal-peoples/Pages/final-report.aspx

Ebert, R. (1997, January 24). Review of Freeway. https://www.rogerebert.com/reviews/freeway-1997

Edin, K., & Shaefer, L.H. (2016). *$2.00 a day: Living on almost nothing in America*. Mariner Books.

Edwards, J. (2014). Oil sands pollutants in traditional foods. *Canadian Medical Association Journal, 186*(12), E444. https://doi.org/10.1503/cmaj.109-4859

Egan, B. (2013). Towards shared ownership: Property, geography, and treaty making in British Columbia. *Geografiska Annaler: Series B, Human Geography, 95*(1), 33–50. https://doi.org/10.1111/geob.12008

Eggerston, L. (2009). High cancer rates among Fort Chipewyan residents. *Canadian Medical Association Journal, 181*(12), E309. https://doi.org/10.1503/cmaj.090248

Eggertson, L. (2015). Canada has 1838 boil water advisories. *Canadian Medical Association Journal, 187*(7), 488. https://doi.org/10.1503/cmaj.109-5018

Ekers, M. (2009). The political ecology of hegemony in depression-era British Columbia, Canada: Masculinities, work and the production of the forestscape. *Geoforum, 40*, 303–15. https://doi.org/10.1016/j.geoforum.2008.09.011

Emery, C. (1992, October). *The northern cod crisis* (Background Paper No. BP-313E). Library of Parliament, Parliamentary Research Branch. https://publications.gc.ca/Collection-R/LoPBdP/BP/bp313-e.htm

Environment and Climate Change Canada. (2021). *Environment and climate change Canada's mandate*. Retrieved February 21, 2023 from https://www.canada.ca/en/environment-climate-change/corporate/mandate.html

Environment Canada. (2011). *Everyone's talking about WATER: It's time for action!* (Catalogue No. En4-119/2011E-PDF). https://ec.gc.ca/Content/D/2/9/D295883B-4FB7-457E-AB8B-9C9D4CA15821/COM1300_Talking_Water_e_web.pdf

Exec. Order No. 12898, 3 C.F.R. 859, 32 C.F.R. § 651 (1994). https://www.govinfo.gov/content/pkg/WCPD-1994-02-14/pdf/WCPD-1994-02-14-Pg276.pdf

Export Development Canada. (2017, March 31). *Canada's forestry sector: Rooted in Canada's economic history*. https://www.edc.ca/en/article/canadas-forestry-sector.html

Fishel, S.R. (2017). *The microbial state: Global thriving and the body politic*. University of Minnesota Press.

Fisheries Act, RSC, 1985, c. F-14, s. 36(3). https://laws-lois.justice.gc.ca/eng/acts/F-14/page-8.html#docCont

Forest, B., & Forest, P. (2012). Engineering the North American waterscape: The high modernist mapping of continental water transfer projects. *Political Geography, 31*(3), 167–83. https://doi.org/10.1016/j.polgeo.2011.11.005

Forest, P. (2010). Inter-local water agreements: Law, geography, and NAFTA. *Les Cahiers de droit, 51*(3–4), 749–70. https://doi.org/10.7202/045732ar

Forest Products Association of Canada. (2019). *Take your place*. Retrieved October 17, 2019 from http://thegreenestworkforce.ca/index.php/take-your-place/

Forestry Stewardship Council. (2020). *Global strategy 2021–2026*. https://fsc.org/sites/default/files/2020-12/FSC%20GLOBAL%20STRATEGY%202021-2026%20%28English%20version%29%20%282%29.pdf

Foster, H. (n.d.). *Aboriginal fisheries in British Columbia*. Indigenous Foundations. Retrieved October 18, 2018 from https://indigenousfoundations.arts.ubc.ca/aboriginal_fisheries_in_british_columbia/

Foucault, M. (1979). *Discipline and punish: The birth of the prison* (A. Sheridan, Trans.). Vintage Books. (Original work published 1977)

Foucault, M. (2009). *Security, territory, population: Lectures at the College de France, 1977–78* (M. Senellart, Ed.; G. Burchell, Trans.). Palgrave Macmillan.

Fraser, N., & Jaeggi, R. (2018). *Capitalism: A conversation in critical theory*. Polity.

Freund, P., & Martin, G. (2000). Driving south: The globalization of auto consumption and its social organization of space. *Capitalism Nature Socialism, 11*(4), 51–71. https://doi.org/10.1080/10455750009358940

Fry, K., & Lousley, C. (2001). Girls just want to have fun with politics: Out of the contradictions of popular culture, eco-grrrls are rising to redefine feminism, environmentalism and political action. *Alternatives Journal, 27*(2), 24–8.

Gest, J. (2018, January 19). Points-based immigration was meant to reduce racial bias. It doesn't. *The Guardian*. https://www.theguardian.com/commentisfree/2018/jan/19/points-based-immigration-racism

Gibson, G., Yung, K., Chisholm, L., & Quinn, H., with Lake Babine Nation, & Nak'azdli Whut'en. (2017). *Indigenous communities and industrial camps: Promoting healthy communities in settings of industrial change*. The Firelight Group. https://firelight.ca/wp-content/uploads/2016/03/Firelight-work-camps-Feb-8-2017_FINAL.pdf

Gibson-Graham, J.K. (2006). *A postcapitalist politics*. University of Minnesota Press.

Gillis, W. (1978). *A history of coal mining in Nova Scotia* (Information Series No. 2). Nova Scotia Department of Mines. https://novascotia.ca/natr/meb/data/pubs/is/is02.pdf

Girvan, A. (2018). *Carbon footprints as cultural-ecological metaphors*. Routledge.

Glacken, C. (1976). *Traces on the Rhodian shore: Nature and culture in Western thought from ancient times to the end of the eighteenth century*. University of California Press.

Gollum, M. (2015, September 19). *Stephen Harper's "old-stock Canadians": Politics of division or simple slip?* CBC News. https://www.cbc.ca/news/politics/old-stock-canadians-stephen-harper-identity-politics-1.3234386

Gomez-Barris, M. (2017). *The extractive zone: Social ecologies and decolonial perspectives*. Duke University Press.

Gosine, A., & Teelucksingh, C. (2008). *Environmental justice and racism in Canada: An introduction*. Emond Publishing.

Government of British Columbia. (n.d.). *Building the railway*. Retrieved February 21, 2023 from https://www2.gov.bc.ca/gov/content/governments/multiculturalism-anti-racism/chinese-legacy-bc/history/building-the-railway

A great destruction. (1996). The History of the Northern Cod Fishery. Retrieved December 12, 2019 from https://www.cdli.ca/cod/histor10.htm

Griffin, K. (2017, May 17). Canada 150: Bill Reid made Haida art recognizable across the country. *Vancouver Sun*. https://vancouversun.com/news/local-news/canada-150/canada-150-bill-reid-made-haida-art-recognizable-across-the-country

Haida Nation v. British Columbia (Minister of Forests), 2004 SCC 73. https://scc-csc.lexum.com/scc-csc/scc-csc/en/item/2189/index.do

Haley, B. (2011). From staples trap to carbon trap: Canada's peculiar form of carbon lock-in. *Studies in Political Economy, 88*(1), 97–132. https://doi.org/10.1080/19187033.2011.11675011

Halpern v. Canada (Attorney General), 2002 CanLII 42749 (ON SCDC). https://canlii.ca/t/7bf5

Hanson, E., & Salomons, T. (n.d.). *Van der Peet case*. Indigenous Foundations. Retrieved February 21, 2023 from https://indigenousfoundations.arts.ubc.ca/van_der_peet_case/

Haraway, D. (2003). *The companion species manifesto: Dogs, people, and significant otherness*. University of Chicago Press.

Hardin, G. (1968). The tragedy of the commons. *Science, 162*(3859), 1243–8. https://doi.org/10.1126/science.162.3859.1243

Harris, C. (2009). *The reluctant land: Society, space, and environment in Canada before Confederation*. UBC Press.

Harris, D.C. (2006). Colonial territoriality: The spatial restructuring of Native land and fisheries on the Pacific coast. In P.R. Sinclair & R.E. Ommer (Eds.), *Power and restructuring: Canada's coastal society and environment* (pp. 35–53). ISER Books.

Harris, L. (1990). *Independent review of the state of the northern cod stock*. Department of Fisheries and Oceans. https://waves-vagues.dfo-mpo.gc.ca/Library/114276.pdf

Harris, R. (2003). From "black-balling" to "marking": The suburban origin of redlining in Canada, 1930s–1950s. *Canadian Geographer, 47*(3), 338–50. https://doi.org/10.1111/1541-0064.00026

Harrison, K. (1996). *Passing the buck: Federalism and Canadian environmental policy*. UBC Press.

Harvey, D. (1982). *The limits to capital*. Basil Blackwell.

Harvey, D. (2001). Globalization and the spatial fix. *Geographische Revue, 3*(2), 23–30. http://geographische-revue.de/archiv/gr2-01.pdf

Harvey, D. (2008). The right to the city. *New Left Review, 53*, 23–40.

Health Canada. (2023). Canada's food guide. Government of Canada. Retrieved January 21, 2023, from https://food-guide.canada.ca/en/

Hern, M., & Johal, A. (with Sacco, J.) (2018). *Global warming and the sweetness of life: A tar sands tale*. MIT Press.

Herod, A., & Wright, M.W. (2002). *Geographies of power: Placing scale*. Blackwell Publishing. https://doi.org/10.1002/9780470773406

Hill, C., & Harrison, K. (2006). Intergovernmental regulation and municipal drinking water. In B. Doern & R. Johnson (Eds.), *Rules, rules, rules: Multilevel regulatory governance in Canada* (pp. 234–58). University of Toronto Press.

Hillman, M., Adams, J., & Whitelegg, J. (1990). *One false move … A study of children's independent mobility*. Policy Studies Institute.

Historica Canada. (n.d.). *Heritage Minutes: Nitro* [Video]. https://www.historicacanada.ca/content/heritage-minutes/nitro

Historica Canada. (1991a). *Heritage Minutes: John Cabot* [Video]. https://www.historicacanada.ca/content/heritage-minutes/john-cabot

Historica Canada. (1991b). *Heritage Minutes: Soddie* [Video]. https://www.historicacanada.ca/content/heritage-minutes/soddie

Historica Canada. (1993). *Heritage Minutes: Maurice Ruddick* [Video]. https://www.historicacanada.ca/content/heritage-minutes/maurice-ruddick

Hoffmann, M. (2019, July 18). *Using language to make the world of fossil fuels strange and ugly*. The Conversation. Retrieved February 21, 2023 from http://theconversation.com/using-language-to-make-the-world-of-fossil-fuels-strange-and-ugly-120204

Holm, W. (Ed.). (1988). *Water and free trade: The Mulroney government's agenda for Canada's most precious resource*. James Lorimer.

Hoogensen, G. (2007). The Canadian fisheries industry: Retrospect and prospect. *Canadian Political Science Review 1*(1), 42–56.

Hornborg, A. (1998). Towards an ecological theory of unequal exchange: Articulating world system theory and ecological economics. *Ecological Economics*, 25(1), 127–36. https://doi.org/10.1016/S0921-8009(97)00100-6

Huang, A. (n.d.). *Cedar*. Indigenous Foundations. Retrieved July 5, 2019 from https://indigenousfoundations.arts.ubc.ca/cedar/

Hunold, C. (2019). Green infrastructure and urban wildlife: Toward a politics of sight. *Humanimalia, 11*(1), 89–109. https://doi.org/10.52537/humanimalia.9479

Idle No More. (n.d.). *An Indigenous-led social movement*. Retrieved February 21, 2023 from https://idlenomore.ca/about-the-movement/

Ilyniak, N. (2014). Mercury poisoning in Grassy Narrows: Environmental injustice, colonialism, and capitalist expansion in Canada. *McGill Sociological Review, 4*, 43–66.

Indian Act, 1985, RSC, c. I-5.

Innis, H.A. (1930). *The fur trade in Canada: An introduction to Canadian economic history*. Yale University Press.

International Commission on Large Dams. (n.d.). *Number of dams by country members*. Retrieved February 7, 2019, from https://www.icold-cigb.org/article/GB/world_register/general_synthesis/number-of-dams-by-country-members

Ireton, J. (2021, September 15). *First Nations want federal party co-operation, commitment to clean water*. CBC News. https://www.cbc.ca/news/canada/ottawa/indigenous-communities-near-ottawa-still-need-clean-water-election-2021-1.6174175

Jackson, J.P., Jr., & Weidman, N.W. (2004). *Race, racism, and science: Social impact and interaction*. ABC-Clio.

Jago, R. (2020, June 30). Canada's national parks are colonial crime scenes. *The Walrus*. https://thewalrus.ca/canadas-national-parks-are-colonial-crime-scenes/

James, R.C., & Mustard, C.A. (2004). Geographic location of commercial plasma donation clinics in the United States, 1980–1995. *American Journal of Public Health, 94*(7), 1224–9. https://doi.org/10.2105/AJPH.94.7.1224

Johnston, J. (1882). General map of part of the North-West Territories including the Province of Manitoba shewing Dominion Land surveys to 31st December 1882, published by order of the Rt. Hon. Sir John A. Macdonald, K.C.B., Minister of the Interior. Additions and corrections to 15th March 1883 [Map]. Library and Archives Canada/e011309128 (Local Class No. H1/701/1883, Box No. 2000229203). https://central.bac-lac.gc.ca/.redirect?app=fonandcol&id=4154165&lang=eng

Jones, E., & Fionda, F. (2017, November 8). In search of Canada's elusive shadow population. *The Discourse*. https://www.thediscourse.ca/data/canadas-shadow-population

Karl, T.L. (1997). *The paradox of plenty: Oil booms and petro-states*. University of California Press.

Keil, R. (2017). *Suburban planet: Making the world urban from the outside in*. Wiley.

Kimmerer, R. (2014). Returning the gift. *Minding Nature, 7*(2), 18–24.

Kines, L. (2019, May 4). Port Renfrew chamber decries logging plan. *Times Colonist*. https://www.timescolonist.com/news/local/port-renfrew-chamber-decries-logging-plan-1.23811698

King, S.J. (2014) *Fishing in contested waters: Place & community in Burnt Church/Esgenoopeititj*. University of Toronto Press.

Kingston, A. (2017, November 22). A bloody mess: The story behind paid plasma in Canada. *Maclean's*. https://www.macleans.ca/society/a-bloody-mess-the-story-behind-paid-plasma-in-canada/

Kondro, W. (1997). Final Krever report paints picture of regulatory disfunction. *The Lancet, 350*(9092), 1688. https://doi.org/10.1016/S0140-6736(05)64294-8

Krever, H. (1997). *Commission of Inquiry on the Blood System in Canada, final report* (4 vols.). https://publications.gc.ca/site/eng/9.698032/publication.html

Kull, C.A., Arnauld de Sartre, X., & Castro-Larrañaga, M. (2015). The political ecology of ecosystem services. *Geoforum, 61*(C), 122–34. https://doi.org/10.1016/j.geoforum.2015.03.004

Laforge, J., & McLachlan, S. (2018). Environmentality on the Canadian prairies: Settler-farmer subjectivities and agri-environmental objects. *Antipode, 50*(2), 359–83. https://doi.org/10.1111/anti.12362

Lamoureux, M. (2015, December 22). The bizarre story of that one time Alberta tried to nuke itself. *Vice Magazine*. https://www.vice.com/en_ca/article/avyqbp/the-bizarre-story-of-that-one-time-alberta-tried-to-nuke-itself

Land Back. (n.d.). *Landback home*. Retrieved December 21, 2022 from https://landback.org/

Larsen, S.C., & Johnson, J.T. (2017). *Being together in place: Indigenous coexistence in a more than human world*. University of Minnesota Press.

Leeson, H. (2002). *Constitutional jurisdiction over health and health care services in Canada* (Discussion Paper No. 12). Commission on the Future of Health Care in Canada. http://publications.gc.ca/collections/Collection/CP32-79-12-2002E.pdf

Lefebvre, H. (1996). *Writings on cities* (E. Kofman & E. Lebas, Eds. and Trans.). Wiley.

Leiss, W. (2011). Modern science, enlightenment, and the domination of nature: No exit? In A. Biro (Ed.), *Critical ecologies: The Frankfurt School and contemporary environmental crises* (pp. 23–42). University of Toronto Press.

Leopold, A. (1966). *A Sand County almanac: With essays on conservation from Round River*. Ballantine Books.

Liboiron, M. (2021). *Pollution is colonialism*. Duke University Press. https://doi.org/10.1215/9781478021445

Linton, J. (2010). *What is water? The history of a modern abstraction*. UBC Press.

Locke, J. (1980). *Second treatise of government* (C.B. MacPherson, Ed.). Hackett. (Original work published 1690)

Loo, T. 2006. *States of nature: Conserving Canada's wildlife in the twentieth century*. UBC Press.

Lothian, W.F. (1987). *A brief history of Canada's national parks*. Environment Canada. http://parkscanadahistory.com/publications/history/lothian/brief/eng/brief-history.pdf

Luke, T.W. (1997). Green consumerism: Ecology and the ruse of recycling. In T. Luke (Ed.), *Ecocritique: Contesting the politics of nature, economy, and culture* (pp. 115–36). University of Minnesota Press.

Maclean's. (2005, November 28). [Cover]. https://web.archive.org/web/20201128023357/https://archive.macleans.ca/issue/20051128

MacDermid, R. (2009). *Funding city politics: Municipal campaign funding and property development in the Greater Toronto Area*. CSJ Foundation for Research and Education and VoteToronto. Retrieved February 21, 2023 from https://www.academia.edu/7457510/Funding_City_Politics

MacNeil, R., & Paterson, M. (2018). Trudeau's Canada and the challenge of decarbonisation. *Environmental Politics, 27*(2), 379–84. https://doi.org/10.1080/09644016.2018.1414747

MacNeil, R., & The Men of the Deeps. (1988). Working Man [Song]. On *Reason to believe*. Virgin Records.

Madden, D., & Marcuse, P. (2016). *In defense of housing: The politics of crisis*. Verso.

Magnusson, W. & Shaw, K. (Eds.). (2002). *A political space: Reading the global through Clayoquot Sound*. University of Minnesota Press.

Maher, S. (2020, January 10). Give us more politicians like John Crosbie, please. *Maclean's*. https://www.macleans.ca/politics/ottawa/give-is-more-politicians-like-john-crosbie-please/

Malm, A. (2016). *Fossil capital: The rise of steam power and the roots of global warming*. Verso.

Maniates, M. (2001). Individualization: Plant a tree, buy a bike, save the world? *Global Environmental Politics, 1*(3), 31–52. https://doi.org/10.1162/152638001316881395

Manning, E. (2000). I am Canadian: Identity, territory and the Canadian national landscape. *Theory & Event, 4*(4). https://muse.jhu.edu/article/32607

Markham, N. (Director). (1994). *Taking stock* [Film]. National Film Board of Canada.

Martin, J.W.G. (2009). *Making settler space: George Dawson, the geological survey of Canada and the colonization of the Canadian West in the Late 19th century* [Doctoral dissertation, Queen's University]. Library and Archives Canada. https://central.bac-lac.gc.ca/.item?id=NR65292&op=pdf&app=Library&oclc_number=771915189

Mauser, W. (2006). Global change in the Anthropocene: Introductory remarks. In E. Ehlers & T. Krafft (Eds.), *Earth system science in the Anthropocene* (pp. 3–4). Springer.

McClenaghan, T., & Lindgren, R. (2018, December 21). Ontario's drinking water rules are not red tape. *Toronto Star*. https://www.thestar.com/opinion/contributors/2018/12/21/ontarios-drinking-water-rules-are-not-red-tape.html

McLuhan, M. (1964). *Understanding media: The extensions of man* (2nd ed.). McGraw-Hill.

Menzies, C.R. (2016). *People of the saltwater: An ethnography of Git Lax M'oon.* University of Nebraska Press.

Menzies, C.R., & Butler, C.F. (2008). The Indigenous foundation of the resource economy of BC's north coast. *Labour/Le Travail, 61*, 131–49.

Merchant, C. (1980). *The death of nature: Women, ecology and the scientific revolution.* Harper & Row.

Millennium Ecosystem Assessment. (2005). *Guide to the Millennium Assessment Reports.* https://www.millenniumassessment.org/en/Index-2.html

Miller, J.R. (2009). *Compact, contract, covenant: Aboriginal treaty-making in Canada.* University of Toronto Press.

Mills, E., & Kalman, H.D. (2007). Architectural history of Indigenous Peoples in Canada. *The Canadian Encyclopedia.* Retrieved July 5, 2019 from https://www.thecanadianencyclopedia.ca/en/article/architectural-history-early-first-nations

Minkow, D. (2017, April 6). *What you need to know about fracking in Canada.* The Narwhal. https://thenarwhal.ca/what-is-fracking-in-canada/

Mirosa, O., & Harris, L.M. (2012). Human right to water: Contemporary challenges and contours of a global debate. *Antipode, 44*(3), 932–49. https://doi.org/10.1111/j.1467-8330.2011.00929.x

Mitchell, T. (2013). *Carbon democracy: Political power in the age of oil.* Verso.

Molson Canada. (2010). *Made from Canada* [Video]. YouTube. https://youtu.be/X_yW4-cgG4g

Monbiot, G. (2018, September 6). We won't save the Earth with a better kind of coffee cup. *The Guardian.* https://www.theguardian.com/commentisfree/2018/sep/06/save-earth-disposable-coffee-cup-green

Moore, J.W. (2015). *Capitalism in the web of life: Ecology and the accumulation of capital.* Verso.

Morgan, G. (2017, December 11). BC to proceed with controversial Site C dam, cost soars to $10.7 billion. *Financial Post.* https://business.financialpost.com/commodities/energy/b-c-to-proceed-with-controversial-site-c-dam-cost-soars-to-10-7-billion

Mortimer-Sandilands, C. (2009). The cultural politics of ecological integrity: Nature and nation in Canada's National Parks, 1885–2000. *International Journal of Canadian Studies, 39–40*, 161–89. https://doi.org/10.7202/040828ar

Mosby, I. (2014). *Food will win the war: The politics, culture, and science of food on Canada's home front.* UBC Press.

Myers, R.A., Hutchings, J.A., & Barrowman, N.J. (1997). Why do fish stocks collapse? The example of cod in Atlantic Canada. *Ecological Applications, 7*(1), 91–106. https://doi.org/10.1890/1051-0761(1997)007[0091:WDFSCT]2.0.CO;2

Natural Resources Canada. (2017). *The state of Canada's forests: Annual report 2017* (Catalogue No. Fo1-6E-PDF). http://cfs.nrcan.gc.ca/pubwarehouse/pdfs/38871.pdf

Nature Conservancy of Canada. (n.d.-a). *Darkwoods*. Retrieved February 21, 2023 from https://www.natureconservancy.ca/en/where-we-work/british-columbia/featured-projects/west-kootenay/darkwoods/
Nature Conservancy of Canada. (n.d.-b) *Darkwoods Forest carbon project*. Retrieved February 21, 2023 from https://www.natureconservancy.ca/en/where-we-work/british-columbia/featured-projects/west-kootenay/darkwoods/dw-carbon.html
Nature Conservancy of Canada. (n.d.-c) *What we do*. Retrieved February 21, 2023 from https://www.natureconservancy.ca/en/what-we-do/
Nelson, J.J. (2008). *Razing Africville: A geography of racism*. University of Toronto Press.
Nikiforuk, A. (2017, April 17). *Canada's methane leakage massively under-reported, studies find*. The Tyee. https://thetyee.ca/News/2017/04/27/Canada-Methane-Leakage-Under-Reported/
Norman, E. (2012). Cultural politics and transboundary resource governance in the Salish Sea. *Water Alternatives*, *5*(1), 138–60.
Nova Scotia Archives. (n.d.). *Nova Scotia mine fatalities, 1838–1992*. Retrieved July 5, 2019 from https://archives.novascotia.ca/meninmines/fatalities/
Nova Scotia Museum of Industry. (n.d.). *Coal mining*. Retrieved July 5, 2019 from https://museumofindustry.novascotia.ca/nova-scotia-industry/coal-mining
O'Brian, J., & White, P. (Eds.). (2007). *Beyond wilderness: The Group of Seven, Canadian identity, and contemporary art*. McGill-Queen's University Press.
O'Connor, D.R. (2002a). *Report of Walkerton Inquiry: Part One, the events of May 2000 and related issues*. Ontario Ministry of the Attorney-General.
O'Connor, D.R. (2002b). *Report of Walkerton Inquiry: Part Two, a strategy for safe drinking water*. Ontario Ministry of the Attorney-General.
O'Connor, J. (1973). *The fiscal crisis of the state*. St. Martin's Press.
Olive, A. (2019). *The Canadian environment in political context* (2nd ed.). University of Toronto Press.
Organisation for Economic Cooperation and Development. (n.d.). *Water withdrawals*. Retrieved June 20, 2020 from https://doi.org/10.1787/4ccfd800-en
Ostrom, E. (1990). *Governing the commons: The evolution of institutions for collective action*. Cambridge University Press.
Paehlke, R. (2004). Environmentalism in one country: Canadian environmental policy in an era of globalization. *Policy Studies Journal*, *28*(1), 160–75. https://doi.org/10.1111/j.1541-0072.2000.tb02021.x
Paquet, M. (2018, February 21). *Canada's merit-based immigration system is no "magic bullet."* The Conversation. https://theconversation.com/canadas-merit-based-immigration-system-is-no-magic-bullet-90923
Parks Canada. (2022). *About the Parks Canada agency*. Government of Canada. Retrieved February 21, 2023 from https://www.pc.gc.ca/en/agence-agency

Parks Canada. (2021). *Parks Canada attendance 2019–20*. Government of Canada. Retrieved February 21, 2023 from https://www.pc.gc.ca/en/docs/pc/attend

Parr, J. (2005). Local water, diversely known. *Environment and Planning D: Society and Space, 23*(2), 251–71. https://doi.org/10.1068/d431

Pasternak, S. (2017). *Grounded authority: The Algonquins of Barriere Lake against the state*. University of Minnesota Press.

Pasternak, S., & Dafnos, T. (2018). How does a settler state secure the circuitry of capital? *Environment and Planning D: Society & Space, 36*(4), 739–57. https://doi.org/10.1177/0263775817713209

Paterson, M. (2007). *Automobile politics: Ecology and cultural political economy*. Cambridge University Press.

Perry, A. (2016). *Aqueduct: Colonialism, resources, and the histories we remember*. ARP Books.

Petro-Canada. (2019). Petro-Canada live by the leaf [Video playlist]. YouTube. Retrieved from February 21, 2023 from https://www.youtube.com/playlist?list=PLWMRezgdS5q7rbElgAOw924-EFunE9xsF

Petrocultures Research Group. (2016). *After oil*. https://afteroil.ca/wp-content/uploads/2022/04/After-Oil.pdf

Pierson, P. (2000). Increasing returns, path dependence, and the study of politics. *American Political Science Review, 94*(2), 251–67. https://doi.org/10.2307/2586011

Port Renfrew Chamber of Commerce. (n.d.). *About Port Renfrew*. Retrieved August 12, 2019 from https://portrenfrewchamber.com/about/about-port-renfrew/

Post, A. (2008, November). *Why trees need fish and fish need trees*. Alaska Fish & Wildlife News. http://www.adfg.alaska.gov/index.cfm?adfg=wildlifenews.view_article&articles_id=407

Princen, T. (2005). *The logic of sufficiency*. MIT Press.

Principles of environmental justice. (1991). https://www.ejnet.org/ej/principles.html

Prudham, S. (2004). Poisoning the well: Neoliberalism and the contamination of municipal water in Walkerton, Ontario. *Geoforum, 35*(3), 343–59. https://doi.org/10.1016/j.geoforum.2003.08.010

Public Health Agency of Canada. (2014). *Supportive environments for physical activity: How the built environment affects our health*. Government of Canada. https://www.canada.ca/en/public-health/services/health-promotion/healthy-living/supportive-environments-physical-activity-built-environment-affects-health.html

Riccioli, J. (2017, October 31). How we got here: Waukesha's journey to get water from Milwaukee almost over. *Milwaukee Journal Sentinel*. https://www.jsonline.com/story/communities/waukesha/news/waukesha/2017/10/31/how-we-got-here-waukeshas-journey-get-water-milwaukee-almost-over/806998001/

Robinson, M. (2013). Veganism and Mi'kmaq legends. *Canadian Journal of Native Studies, 33*(1), 189–96. https://doi.org/10.4324/9781003013891-4

Rogers, H.R. (1943). Attack on all fronts (Box No. A538/X3, ID No. 2860024). Library and Archives Canada. https://central.bac-lac.gc.ca/.redirect?app=fonandcol&id=2860024&lang=eng

Ross, S. (2018, February 9). A study in news poverty in Thunder Bay. *J-Source*. https://j-source.ca/article/study-news-poverty-thunder-bay/

Rossiter, D. (2004). The nature of protest: Constructing the spaces of British Columbia's rainforests. *Cultural Geographies, 11*(2), 139–64. https://doi.org/10.1191/14744744004eu298oa

Rozworski, M. (2018, November 1). Are we addicted to debt? Bubble or no bubble, the housing affordability crisis creates systemic risk for the Canadian economy. *The Monitor*. https://www.policyalternatives.ca/publications/monitor/are-we-adicted-debt

Rudin, R. (2011). The First French-Canadian national parks: Kouchibouguac and Forillon in history and memory. *Journal of the Canadian Historical Association, 22*(1), 161–200. https://doi.org/10.7202/1008961ar

Rustad, H. (2016, September 19). Big Lonely Doug: How a single tree, and the logger who saved it, have changed the way we see British Columbia's old-growth forests. *The Walrus*. https://thewalrus.ca/big-lonely-doug/

Ryan, S. (1990, March 26). *History of Newfoundland cod fishery*. The History of the Northern Cod Fishery. Retrieved December 12, 2019 from https://www.cdli.ca/cod/history5.htm

Safe Drinking Water Act, 2002, SO 2002, c. 32. https://www.ontario.ca/laws/statute/02s32

Salomons, T. (n.d.). *Calder case*. Indigenous Foundations. Retrieved February 21, 2023 from https://indigenousfoundations.arts.ubc.ca/calder_case/

Salomons, T., & Hanson, E. (n.d.). *Sparrow case*. Indigenous Foundations. Retrieved February 21, 2023 from https://indigenousfoundations.arts.ubc.ca/sparrow_case/

Sandilands, C. (1998). The good-natured feminist: Ecofeminism and democracy. In R. Keil, D.V.J. Bell, P. Penz, & L. Fawcett (Eds.), *Political ecology: Global and local* (pp. 237–49). Routledge.

Schlosberg, D. (2009). *Defining environmental justice: Theories, movements, and nature*. Oxford University Press.

Scott, A. (1982). Regulation and the location of jurisdictional powers: The fishery. *Osgoode Hall Law Journal, 20*(4): 780–805. https://digitalcommons.osgoode.yorku.ca/ohlj/vol20/iss4/6

Scott, J. (1998). *Seeing like a state: How certain schemes to improve the human condition have failed*. Yale University Press.

Shaefer, H.L., & Ochoa, A. (2018, March 15). How blood-plasma companies target the poorest Americans: The industry's business model depends on there being plenty of people who need cash quickly. *The Atlantic*. https://www.theatlantic.com/business/archive/2018/03/plasma-donations/555599/

Siemiatycki, M. (2016). Public private partnerships in Canada. In A. Akintoye, M. Beck, & M. Kumaraswamy (Eds.), *Public private partnerships: A global review* (pp. 59–73). Routledge.

Simone, D., & Walks, A. (2019). Immigration, race, mortgage lending, and the geography of debt in Canada's global cities. *Geoforum, 98*, 286–99. https://doi.org/10.1016/j.geoforum.2017.10.006

Simpson, A. (2014). *Mohawk interruptus: Political life across the borders of settler states*. Duke University Press.

Simpson, L.B. (2014). Land as pedagogy: Nishnaabeg intelligence and rebellious transformation. *Decolonization: Indigeneity, Education & Society, 3*(3), 1–25.

Simpson, L.B. (2016). How to steal a canoe [Video]. Vimeo. https://vimeo.com/188380371

Simpson, L.B. (2017). *As we have always done*. University of Minnesota Press.

Simpson, L., DaSilva, J., Riffel, B., & Sellers, P. (2009). The responsibilities of women: Confronting environmental contamination in the traditional territories of Asubpeechoseewagong Netum Anishinabek (Grassy Narrows) and Wabauskang First Nation. *International Journal of Indigenous Health, 4*(2), 6–13.

Smith, N. (2019, September 9). *Fracking is the bridge to renewable energy: Banning it now will only make the goal of a green America harder to achieve*. Bloomberg. https://www.bloomberg.com/opinion/articles/2019-09-09/fracking-is-the-bridge-to-a-fully-renewable-future

Smith, R.F. (1978). History and current status of irrigation in Alberta. *Canadian Water Resources Journal, 3*(1), 6–18. https://doi.org/10.4296/cwrj0301006

Sneddon, C., Harris, L., Dimitrov, R., & Özesmi, U. (2002). Contested waters: Conflict, scale, and sustainability in aquatic socio-ecological systems. *Society and Natural Resources, 15*(8), 663–75. https://doi.org/10.1080/08941920290069272

Sofoulis, Z. (2005). Big water, everyday water: A sociotechnical perspective. *Continuum, 19*(4), 445–63. https://doi.org/10.1080/10304310500322685

Soper, K. (1995). *What is nature? Culture, politics, and the non-human*. Blackwell.

Species at Risk Act, SC 2002, c. 29. https://laws.justice.gc.ca/eng/acts/s-15.3/

Springer, A.L. (1997). The Canadian turbot war with Spain: Unilateral state action in defense of environmental interests. *Journal of Environment & Development, 6*(1), 26–60. https://doi.org/10.1177/107049659700600103

Stanford, J. (2008). *Economics for everyone: A short guide to the economics of capitalism*. Fernwood.

Statistics Canada. (2017, June 27). *150 years of Canadian agriculture* (Catalogue No. 11-627-M). https://www150.statcan.gc.ca/n1/pub/11-627-m/11-627-m2017018-eng.pdf

Steffen, W., Crutzen, P., & McNeill, J. (2007). The Anthropocene: Are humans now overwhelming the great forces of nature? *Ambio, 36*(8), 614–21. https://doi.org/10.1579/0044-7447(2007)36[614:TAAHNO]2.0.CO;2

Steffen, W., Persson, A., Deutsch, L., Zalasiewicz, J., Williams, M., Richardson, K., Crumley, C., Crutzen, P., Folke, C., Gordon, L., Molina, M., Ramanathan, V., Rockström, J., Scheffer, M., Schellnhuber, H.J., & Svedin, U. (2011). The Anthropocene: From global change to planetary stewardship. *Ambio, 40*(7), 739–61. https://doi.org/10.1007/s13280-011-0185-x

St. John, M. (Director). (2016). *Colonization road* [Film]. Decolonization Roads Production Inc.

Sultana, F., & Loftus, A. (2015). The human right to water: Critiques and condition of possibility. *WIREs Water, 2*(2), 97–105. https://doi.org/10.1002/wat2.1067

Sustainable Human. (2014, February 13). How wolves change rivers. Narr. G. Monbiot. YouTube. https://youtu.be/ysa5OBhXz-Q

Swardson, A. (1995, March 29). Canada's fish affair: Diplomacy or piracy? *The Washington Post*. https://www.washingtonpost.com/archive/politics/1995/03/29/canadas-fish-affair-diplomacy-or-piracy/1b1df60f-5a2a-4d95-adcb-01dd815f7e14/

Teelucksingh, C., Poland, B., Buse., C, & Hasdell, R. (2016). Environmental justice in the environmental non-governmental organization landscape of Toronto (Canada). *Canadian Geographer, 60*(30), 381–93. https://doi.org/10.1111/cag.12278

Theckedath, D., & Thomas, T.J. (2012). *Media ownership and convergence in Canada* (Publication No. 2012-17-E). Library of Parliament. Retrieved May 15, 2020 from https://publications.gc.ca/collections/collection_2012/bdp-lop/eb/2012-17-eng.pdf

Truth and Reconciliation Commission of Canada. (2015). Truth and Reconciliation Commission of Canada: Calls to action. https://ehprnh2mwo3.exactdn.com/wp-content/uploads/2021/01/Calls_to_Action_English2.pdf

Tsilhqot'in Nation v. British Columbia, 2014 SCC 44. https://scc-csc.lexum.com/scc-csc/scc-csc/en/item/14246/index.do

Tuck, E., & Wang, K.W. (2012). Decolonization is not a metaphor. *Decolonization: Indigeneity, Education & Society, 1*(1), 1–40.

Turner, J.M. (2002). From woodcraft to "leave no trace": Wilderness, consumerism, and environmentalism in twentieth-century America. *Environmental History, 7*(3), 462–84. https://doi.org/10.2307/3985918

Van Praet, N. (2019, May 23). Quebec Oil and Gas Association rebrands as Quebec Energy Corporation. *The Globe and Mail*. https://www.theglobeandmail.com/business/article-quebec-oil-and-gas-association-rebrands-as-quebec-energy-corporation/

Venton, M., & Mitchell, K. (2015, January 29). *A dark day for environmental justice in Canada*. Huffington Post. https://www.huffingtonpost.ca/ecojustice/environmental-justice-canada_b_6547682.html?guccounter=1

Vincent, D. (2020a, May 7). Sidewalk Labs pulls out of Toronto's Quayside project, blaming COVID-19. *Toronto Star*. https://www.thestar.com/news/city_hall/2020/05/07/sidewalk-labs-pulling-out-of-quayside-project.html

Vincent, D. (2020b, February 27). Waterfront Toronto advisory panel still has concerns about Sidewalk Labs' data collection, new report says. *Toronto Star*. https://www.thestar.com/news/gta/2020/02/26/waterfront-toronto-advisory-panel-still-has-concerns-about-sidewalk-labs-data-collection-new-report-says.html

Waldron, I. (2018). *There's something in the water: Environmental racism in Indigenous and Black communities*. Fernwood.

Walters, D., Spence, N., Kuikman, K., & Singh, B. (2012). Multi-barrier protection of drinking water systems: A comparison of First Nations and non-First Nations communities in Ontario. *International Indigenous Policy Journal, 3*(3), 1–25. https://doi.org/10.18584/iipj.2012.3.3.8

Water Canada. (2016). $250 billion: Sink or swim: How can Canada finance the infrastructure gap. *Water Canada, 16*(5).

Watkins, M. (1963). A staple theory of economic growth. *Canadian Journal of Economics and Political Science, 29*(2), 49–73. https://doi.org/10.2307/139461

Weaver, K. (2015, November). The Lost Town of Pine Point. *Up Here*. https://uphere.ca/articles/lost-town-pine-point

Weber, L. (2018). *Sydney tar ponds contamination, Nova Scotia Canada*. Environmental Justice Atlas. Retrieved February 21, 2023 from https://ejatlas.org/conflict/sydney-tar-ponds-contamination-nova-scotia-canada

Weibe, S. (2017). *Everyday exposure: Indigenous mobilization and environmental justice in Canada's Chemical Valley*. UBC Press

Weibust, I. (2009). *Green leviathan: The case for a federal role in environmental policy*. Ashgate.

Weldon, J. (Director). (1979). *Log driver's waltz* [Film]. National Film Board of Canada. https://www.nfb.ca/film/log_drivers_waltz/

Wilson, A. (1991). *The culture of nature: North American landscape from Disney to the Exxon Valdez*. Between the Lines.

Wilson, G.N., & Bowles, P. (2015). *Resource communities in a globalizing region: Development, agency, and contestation in Northern British Columbia*. UBC Press.

Wilson, K. (2007). The Krever Commission: 10 years later. *Canadian Medical Association Journal, 177*(11), 1387–9. https://doi.org/10.1503/cmaj.071333

Wolfe, P. (2006). Settler colonialism and the elimination of the native. *Journal of Genocide Research, 8*(4), 387–409. https://doi.org/10.1080/14623520601056240

World Commission on Dams. (2000). *Dams and development: A new framework for decision-making*. Earthscan.

Worster, D. (1985). *Rivers of empire*. Oxford University Press.

Wright, M. (2008). Building the great lucrative fishing industry: Aboriginal gillnet fishers and protests over salmon fishery regulations for the Nass and Skeena Rivers, 1950s–1960s. *Labour/Le Travail, 61*(Spring), 99–130. https://www.jstor.org/stable/25149856

Wynn, G. (2006). Foreword. In T. Loo, *States of nature: Conserving Canada's wildlife in the twentieth century* (pp. xi–xxi). UBC Press.

Yusoff, K. (2013). Geologic life: Prehistory, climate, futures in the Anthropocene. *Environment and Planning D: Society and Space, 31*(5), 779–95. https://doi.org/10.1068/d11512

Zuboff, S. (2019). *The age of surveillance capitalism: The fight for a human future at the new frontier of power*. Public Affairs.

Index

Notes: The letter *f* following a page number denotes a figure; the letter *g*, a term found within the glossary; the letter *m*, a map; the letter *t*, a table.

Printed and bound by CPI Group (UK) Ltd, Croydon, CR0 4YY

09/06/2025

14685785-0002